Transformative Teaching

A Collection of Stories of
Engineering Faculty's Pedagogical Journeys

Synthesis Lectures on Engineering

Each book in the series is written by a well known expert in the field. Most titles cover subjects such as professional development, education, and study skills, as well as basic introductory undergraduate material and other topics appropriate for a broader and less technical audience. In addition, the series includes several titles written on very specific topics not covered elsewhere in the Synthesis Digital Library.

Transformative Teaching: A Collection of Stories of Engineering Faculty's Pedagogical Journeys
Nadia Kellam, Brooke Coley, and Audrey Boklage
2019

Ancient Hindu Science: Its Transmission and Impact of World Cultures
Alok Kumar
2019

Value Relational Engineering
Shuichi Fukuda
2018

Strategic Cost Fundamentals: for Designers, Engineers, Technologists, Estimators, Project Managers, and Financial Analysts
Robert C. Creese
2018

Concise Introduction to Cement Chemistry and Manufacturing
Tadele Assefa Aragaw
2018

Data Mining and Market Intelligence: Implications for Decision Making
Mustapha Akinkunmi
2018

Empowering Professional Teaching in Engineering: Sustaining the Scholarship of Teaching
John Heywood
2018

The Human Side of Engineering
John Heywood
2017

Geometric Programming for Design Equation Development and Cost/Profit
Optimization (with illustrative case study problems and solutions), Third Edition
Robert C. Creese
2016

Engineering Principles in Everyday Life for Non-Engineers
Saeed Benjamin Niku
2016

A, B, See... in 3D: A Workbook to Improve 3-D Visualization Skills
Dan G. Dimitriu
2015

The Captains of Energy: Systems Dynamics from an Energy Perspective
Vincent C. Prantil and Timothy Decker
2015

Lying by Approximation: The Truth about Finite Element Analysis
Vincent C. Prantil, Christopher Papadopoulos, and Paul D. Gessler
2013

Simplified Models for Assessing Heat and Mass Transfer in Evaporative Towers
Alessandra De Angelis, Onorio Saro, Giulio Lorenzini, Stefano D'Elia, and Marco Medici
2013

The Engineering Design Challenge: A Creative Process
Charles W. Dolan
2013

The Making of Green Engineers: Sustainable Development and the Hybrid Imagination
Andrew Jamison
2013

Crafting Your Research Future: A Guide to Successful Master's and Ph.D. Degrees in
Science & Engineering
Charles X. Ling and Qiang Yang
2012

Fundamentals of Engineering Economics and Decision Analysis
David L. Whitman and Ronald E. Terry
2012

A Little Book on Teaching: A Beginner's Guide for Educators of Engineering and
Applied Science
Steven F. Barrett
2012

Engineering Thermodynamics and 21st Century Energy Problems: A Textbook
Companion for Student Engagement
Donna Riley
2011

MATLAB for Engineering and the Life Sciences
Joseph V. Tranquillo
2011

Systems Engineering: Building Successful Systems
Howard Eisner
2011

Fin Shape Thermal Optimization Using Bejan's Constructal Theory
Giulio Lorenzini, Simone Moretti, and Alessandra Conti
2011

Geometric Programming for Design and Cost Optimization (with illustrative case study
problems and solutions), Second Edition
Robert C. Creese
2010

Survive and Thrive: A Guide for Untenured Faculty
Wendy C. Crone
2010

Geometric Programming for Design and Cost Optimization (with Illustrative Case Study
Problems and Solutions)
Robert C. Creese
2009

Style and Ethics of Communication in Science and Engineering
Jay D. Humphrey and Jeffrey W. Holmes
2008

Introduction to Engineering: A Starter's Guide with Hands-On Analog Multimedia
Explorations
Lina J. Karam and Naji Mounsef
2008

Transformative Teaching: A Collection of Stories of Engineering Faculty's Pedagogical Journeys

Nadia Kellam, Brooke Coley, and Audrey Boklage

ISBN: 978-3-031-79405-6   paperback
ISBN: 978-3-031-79406-3   ebook
ISBN: 978-3-031-00149-9   hardcover

DOI 10.1007/978-3-031-79406-3

A Publication in the Springer series
SYNTHESIS LECTURES ON ENGINEERING

Lecture #35
Series ISSN
Print 1939-5221   Electronic 1939-523X

Cover image by Pexels on Pixabay (https://pixabay.com/users/Pexels-2286921/).
Image retrived from https://pixabay.com/photos/mountains-dawn-dusk-grass-hills-1868715/

Transformative Teaching

A Collection of Stories of
Engineering Faculty's Pedagogical Journeys

Nadia Kellam
Arizona State University

Brooke Coley
Arizona State University

Audrey Boklage
University of Texas at Austin

SYNTHESIS LECTURES ON ENGINEERING #35

ABSTRACT

The journey to becoming an exemplary engineering educator is one that is rarely simple and straightforward. Simply being exposed to active learning strategies or innovative pedagogies rarely leads to a transformation of one's own teaching. In this book, we present a collection of stories from exemplary engineering educators that are told in their own voices. These stories are shared to enable readers to immerse themselves in first-person recollections of transformation, involving engineering educators who changed their teaching strategies from the ways that they were taught as engineering undergraduate students to ways that more effectively fostered a conducive learning atmosphere for all students. It is our hope that providing stories of successful engineering educators might stimulate thoughtful and productive self-reflection on ways that we can each change our own teaching. These stories are not simple, linear stories of transformation. Instead, they highlight the complexities and nuances inherent to transforming the way that engineering faculty teach. Through our strategy of narrative storytelling, we hope to inspire future and current engineering educators to embark on their own journeys of teaching transformations. We conclude the book with some lessons that we learned during our readings of these stories, and invite readers to extract lessons of their own.

KEYWORDS

engineering teaching journeys, engineering teaching stories, innovative engineering teaching, engineering active learning, narrative interviews of engineering faculty, exemplary engineering teachers, exemplary engineering educators, innovative engineering teachers, innovative engineering educators, innovation in engineering education, engineering teaching inspiration, engineering educator inspiration, engineering faculty teaching stories

Contents

Preface

When I was interviewing for my first engineering tenure-track faculty position in 2006, the department head asked me which classes I could teach in the program. Confidently, I responded that I could teach anything I had taken. A few months later, in the summer before I started, I learned I would be teaching something I had never taken before—a computational engineering methods class for first-year engineering students. After talking to a few people, I understood that the undergraduate program coordinator wanted to get someone into the classroom who was nice to first-year students. The current teacher, an Associate Professor, had structured the entire class around programming and was known to get frustrated with the students with rumors of him throwing chalk at students. I could be nice, I thought, but I'm not sure about teaching the course with my degrees in mechanical engineering. I managed to get by through changing the course to focus on learning programming structures (through storytelling using Alice), learning relative and absolute referencing in Microsoft Excel, learning html to create websites, and learning basic programming using MatLAB. I ended up teaching a few sections of this class for two years and eventually became comfortable teaching computational engineering methods to first-year students. I even ended up having some fun with the class with a challenge at the end of the course for the students to use MatLAB to create artwork for an art exhibit. Students created music, edited photographs, created fractals, and created stop motion animations. During the art exhibit, I overheard one of my senior colleagues make a comment to the students saying that this was not engineering and seemed like a waste of time to him. This was my first experience teaching as an engineering faculty member.

After that first semester teaching my own class, I quickly realized how difficult and complicated teaching could be. While I had read many books preparing me to be a teacher, it was hard to truly prepare for experiences like this. I did not expect to be asked to teach something that I had not learned myself as a student. I also expected that senior faculty members in the department would be supportive of a junior faculty member. In hindsight, I was a bit naïve and probably should not have been surprised to experience some pushback for my alternative ways of teaching. I was, after all, the second woman faculty in our department of around 50 faculty members. In addition, I was the youngest faculty. The composition of our faculty was about to change, but when I first joined it was pretty homogeneous.

Because the teaching books that I read did not seem to be helping prepare me for the realities of teaching, I sought out additional opportunities to develop my teaching skills. These included workshops such as the National Effective Teaching Institute (NETI) that Rich Felder and Rebecca Brent hosted before the American Society for Engineering Education (ASEE) conference. However, in spite of these intensive experiences to help me as a teacher, it felt difficult to reconcile the actual experience of teaching with what I was learning from the experts. I would learn helpful strategies for teaching, but some of the difficulties I faced were not addressed in these workshops. In this book, we are going to share the messy and sometimes complicated stories of faculty as they embark on journeys to become better teachers. I hope that, through immersing yourself in these stories, that you will learn more about the journeys that faculty take to become better teachers and feel better prepared as you embark on your own journey.

Nadia Kellam
April 2019

Acknowledgments

We wish to thank our fellow research team members, Joachim Walther, Stephan Durham, Sandra Bird, and Kathleen DeMarrais at the University of Georgia for support during the early stages of this project. We would also like to thank more recent research team members including Madeleine Jennings, Joshua Cruz, Michael Sheppard, and Anna Cirell at Arizona State University for their support during the later stages of this book preparation. We would especially like to thank Madeleine Jennings for helping with copy-editing many of the chapters of this book. In addition, we would like to thank the research participants in this study, including those whose stories were not included in this book.

This material is based upon work supported by the National Science Foundation under grant numbers 1329300 and 1542531. Any opinions, findings, and conclusions or recommendations expressed in this material are those of the authors and do not necessarily reflect the views of the National Science Foundation.

Nadia Kellam, Brooke Coley, and Audrey Boklage
April 2019

CHAPTER 1

Introduction

Nadia Kellam, Audrey Boklage, and Brooke Coley

Anyone who has been an undergraduate engineering student knows that exemplary engineering educators are rare to find but can make a big difference in engineering students' development, sometimes being the difference between persisting in engineering or changing majors. As some of us continue through school and then become engineering faculty members, it can be daunting to figure out how to become good teachers ourselves, especially when our own Master's and Ph.D. programs tend to focus on research and engineering sciences with little, if any, focus on our development as teachers. In this book, we (in this chapter, "we" refers to Nadia, Audrey, and Brooke) hope to provide an opportunity for engineering educators to learn about engineering teaching through becoming immersed in the stories of exemplary engineering faculty. These are not polished and overly edited stories of engineering faculty, but instead are somewhat raw and uncut stories as told by the faculty themselves. These stories were developed from transcriptions of narrative interviews and are kept in the spoken words of the engineering faculty. We wanted to explore ways of sharing the experience of hearing these inspirational stories, and thus the concept of this book was born. These stories promise to humanize teachers and show that good teachers are not just "born that way," but face many obstacles on their own journeys of becoming exemplary teachers.

From the outset of this project, we were interested in using narrative research strategies (narrative interviews and analysis), to develop an understanding of the stories of successful engineering faculty who have embraced active learning strategies [Meyers and Jones, 1993]. Stories are powerful ways of learning from others and are inherent to the way that we communicate, think, and learn. "We think in story. It's hardwired in our brain. It's how we make strategic sense of the otherwise overwhelming world around us" [Cron, 2012, p. 8]. We wholeheartedly agree that people are "hardwired" for stories and hope that by sharing these faculty stories others can be inspired and better prepared to embark on their own teaching journeys.

As you immerse yourself into the stories shared in this book, in addition to providing inspiration, we hope the stories resonate with you and help you as you navigate your own personal teaching journeys. Through experiencing these stories, we can all learn from these stories, and, possibly, see teaching in a different way than we did before. I know for me (in this chapter, "I" refers to Nadia), they impacted my understanding of what it means to be a good teacher. Somewhere in the back of my mind, I thought there were some teachers who were simply born amazing teachers. From my experience, I knew I was not one of these fortunate people but hoped that with enough work I could become better. Through these stories, I began to see that other's journeys are not simple and that even amazing teachers experience trials and difficulties throughout their journeys.

MOTIVATION FOR THIS PROJECT

When reading literature about STEM or engineering faculty and how they become exemplar teachers, much focuses on why faculty do not embrace active learning strategies. For example, in the Discipline Based Educational Research report, the committee describes barriers to changing teaching practices including institutional priorities, local leadership, peers, reward systems, students' attitudes, perceived importance of teaching, and faculty members' beliefs [National Research Council, 2012]. While we recognize the importance of identifying and understanding these barriers, we are also interested in understanding faculty who have successfully changed their teaching practices. We decided to focus this research on these faculty who have successfully transitioned to active learning strategies and to uncover insights and lessons learned from their stories.

The interviews that were used as the basis for these stories came from a research project focused on engineering faculty change. When we conducted these interviews, we did so for the purpose of the research project, and not with the goal of writing a book that included their stories. However, as we began conducting interviews, we quickly became inspired by the stories of the interviewees. Many of the interviews were conducted by Brooke and Audrey, who were, at the time of the data collection, postdoctoral researchers. In our team meetings after interviews, they were both very excited about the stories that they were hearing. We began to see the power of hearing the inspirational stories of these engineering faculty. In addition, I, who had been a faculty member for about 10 years, would listen to the interviews and become just as inspired and excited as Brooke and Audrey. These stories were powerful and we began to consider ways that these could be shared in a more complete form so that more people could become inspired and empowered by these stories.

Another observation when conducting interviews for this research project was that the stories of faculty change were complex stories. They were not stories of faculty who just happened to be amazing teachers from day one. Nor were they stories of faculty who decided to make a change to their teaching, made their change, and succeeded easily. Instead, they were stories of faculty who wanted to make changes to their teaching for different reasons, and they all encountered successes and struggles. In other words, these were not simple or linear stories of change. Instead, they were messy and complicated stories of change. In a few of the cases, these stories had reached a conclusion, and in others, the journey was ongoing and the engineering faculty members were continuously evolving as teachers.

In Chapters 2–9 of this book, we will present these stories of faculty who have successfully transitioned to active learning strategies in their classes. These stories were developed based on interviews with these faculty and are kept in their words as spoken.

HOW WE STRUCTURED THESE STORIES

As described above, these stories were captured as part of a research project where we interviewed exemplar teachers to develop an understanding of how they got to where they are today in spite of all the challenges and obstacles along the way. As part of our methods, we constructed stories, in the spoken word of the participants, as we felt this format had the most resonance with the reader. We used Joseph Campbell's *Hero's Journey* [2008] as a way to structure these stories. We then analyzed the data for patterns across the stories. However, when we disseminated this work in journal articles, most of the participants' stories and voices were lost. Their stories were reduced to a few pages, at most, with a few supporting quotes from their interviews [Boklage, Coley, and Kellam, 2018].

Because we were unable to share these complete stories in traditional dissemination venues, we began considering nontraditional ways of sharing these stories. After some consideration, we decided to prepare a book that would include faculty stories in their entirety. The hope is that these stories will serve as an inspiration to help teachers as they embark on or grow in their own personal journeys of transformation.

Prior to sharing these stories, we will describe the Hero's Journey, as the stories in this book are all organized using this structure. The idea behind the Hero's Journey is that all stories follow similar structural patterns. In Joseph Campbell's book, *Hero with a Thousand Faces*, he introduces the monomyth [2008]. The monomyth is a universal structure that all epic myths are claimed to follow. In Campbell's book, he considers over 100 stories from multiple cultures and times and shows that these stories follow a similar trajectory. Campbell proposes 17 stages that stories generally follow. We have interpreted these stages that were intended for written or told stories or epic myths for lived stories of engineering faculty. Below are brief descriptions of the stages that we used in structuring stories that were told in interviews.

1. The *call to adventure* marks the beginning of the faculty's story and includes their purpose or reason for embarking on a journey.

2. The *refusal of the call* occurs after the call to adventure and involves the faculty member changing their mind and deciding to not begin their journey. This is typically a consideration and, at least in the stories of faculty who have successfully transitioned their teaching practices highlighted in this book, is only a consideration that does not result in the end of the journey.

3. *Supernatural aid* occurs when the faculty member receives unexpected help from a mentor, colleague, or other resource (e.g., a book or website). This aid helps the faculty member prepare for the journey that they are about to take.

4. *The first threshold* is experienced when the faculty member continues forward in their journey and experiences their first trial or challenge on their journey. This challenge is typically expected by the faculty member, as they anticipate some difficulties when embarking on the journey.

5. During the *belly of the whale*, the faculty member experiences a very low point in their journey. Oftentimes, this experience becomes transformative for the faculty member as they have a realization of the importance of this journey as they recover from this low point.

6. During the *road of trials*, the faculty member experiences and overcomes many challenges. This could be student resistance to active learning strategies, or colleagues questioning the effort being put into teaching.

7. The *meeting with the all-knower* structure represents the faculty member meeting with a mentor who passes critical knowledge onto the faculty member. Without this interaction, it could be imagined that the journey might have ended very differently for the faculty member.

8. The *meeting with temptation* occurs when the faculty member has an experience that could keep them from reaching their personal goal. This temptation could be in the form of focusing efforts on research instead of teaching, beginning to lecture again because of the potential of earning higher teaching evaluations, or following traditional course approaches after some students express frustration with the new approaches.

9. In the *apotheosis* stage, the faculty member reaches a new level of understanding where their journey becomes routine and their teaching innovations come with fewer surprises.

10. The *ultimate boon* occurs near the end of the journey as the journey reaches resolution. As could be imagined, many of the faculty in this book do not reach the end of their story, but do reach some boon where they attain a steady state in their goals.

11. The *return threshold* occurs when the faculty member begins communicating with peers, colleagues, students, and administrators, telling them what they learned during their journey and beginning to reconcile their new identity with the one they left behind as they embarked on their journey.

12. The final phase, *master of both worlds and freedom to live*, represents when the faculty member moves back to the "ordinary" world that they left when embarking on their journey to the "special" world that they inhabited while on their journey. This can involve sharing their story of change with people who have not embarked on their own journey. It can also involve becoming integrated back into the "ordinary" world with the knowledge gained while on their journey.

There are five additional stages that were not used in these journeys, and will not be explained in detail. These include some that are less applicable to lived stories, including, for example, the *magic flight* which involved the hero rushing home in a pursuit. Others were just not included in the stories highlighted in this book and include, for example, the *refusal of the return*, where a faculty member would refuse to move back into their "ordinary" world after experiencing the "special" world.

As you begin reading the chapters, you will notice that many of the stories only include some of the stages in the journey. These stages were only used to structure the stories as they were constructed from the spoken interview. These stages will be used to help organize the subsequent chapters. For those interested in journal articles, we outline this process in Cruz and Kellam [2017] and use this structure in an article exploring the beginning of engineering students' journeys [Cruz and Kellam, 2018].

In addition to the stages described above, we added some stages to the stories. One common addition is named *stories from my class*. While the monomyth provided a helpful structure for organizing and constructing narratives from the interview transcripts, we found that some parts of the story were excluded because they did not follow neatly into one of the structures. While the *stories from my class* structure did not involve a particular trial or challenge, we felt it was important to include this part of their story as it showed innovations in their teaching and provided more texture and context to their particular journey.

In each participant's story we will use headings to denote each stage in the journey. This will help the reader move more easily between stories to compare, for example, specific stages for each engineering educator.

OVERVIEW OF THE BOOK

The participants' stories are told in their spoken voice as transcribed from the interview. By keeping the stories true to their voice, we believe that the stories are more engaging than they would be if we rephrased them. This does mean that there are some run-on sentences and colloquial terms used in their stories. Occasionally, we include a few additional words to help improve the flow of the story. These words are denoted with square brackets in the text. In addition, we provide some clarifying details in parentheses (e.g., the meaning of an acronym).

In Chapter 2, Donna Riley, the Kamyar Haghighi Head of the School of Engineering Education at Purdue University, shares her teaching story. At the time of the interview, Donna was a faculty member at Virginia Tech. Donna tells her story of integrating a liberative pedagogy into engineering education. After she started her first faculty appointment at Smith College, she began a 10-year experiment in a Thermodynamics course where she challenged the power dynamics common to engineering courses and pushed students to begin thinking critically about the subject. Her story is one that includes social activism and is one that will serve as an inspiration to many faculty as she challenged the status quo in engineering education.

In Chapter 3, Sara Atwood, an Associate Professor and Chair of Engineering and Physics at Elizabethtown College, shares her teaching story. Her undergraduate studies at Dartmouth College, a liberal arts setting, provided her foundation for student-centered learning. Sara's journey was one that elevated the evolution, process, and development of implementing this pedagogical approach. Among her main supports were the colleagues and community created around these efforts, like-minds committed to enhancing the education of engineering students.

In Chapter 4, Brad Hyatt, an Associate Professor of Construction Management at Fresno State University, shares his story. Prior to being a faculty member, Brad worked for 12 years in industry, both as a Civil Engineering Officer in the Navy and as a Project Management and Construction Management Consultant. When he was a new faculty member, Brad approached teaching with a lot of energy and a "just do it" attitude where he adopted project-based learning, flipping the classroom, and bringing case studies into the class.

In Chapter 5, Chris Swan, an Associate Professor in Civil and Environmental Engineering at Tufts University, shares his story. His belief is that students should experience knowledge and he works to connect content with applications. He finds seeing the application in a real-world context to be especially critical in students' ability to truly grasp material and he facilitates this by offering students service-learning based projects.

In Chapter 6, Thais Alves, an Associate Professor of Construction Engineering at San Diego State University, shares her story. Thais brings an international experience as she is from Brazil, completed her Ph.D. at UC Berkeley, and returned to Brazil again prior to becoming a faculty member in San Diego. When Thais became a faculty member in San Diego, she had to become creative with her teaching because she did not have the access to construction sites that she had in Brazil. She began to take an entrepreneurial approach to her teaching and considered her students as clients to find a way that students begin to value what they were learning in class. Now, she integrates site visits, food, and Lego simulations into her classes.

In Chapter 7, Fernanda Leite, an Associate Professor in Civil, Architectural, and Environmental Engineering in the Cockrell School of Engineering at The University of Texas in Austin, shares her story. Throughout her experiences, Fernanda has always been passionate about teaching and as a graduate student she revamped a lab course while she worked as a Teaching Assistant (TA). At UT Austin, Fernanda has developed courses where she created modules that connect lectures, lab classes, and reflections across topics in the course. She brings real-world scenarios into the classroom where students have to make assumptions and estimates. She also discusses how her teaching and research have been inseparable with each one enhancing the other.

In Chapter 8, Charles Pierce, an Associate Professor of Civil and Environmental Engineering at the University of South Carolina, shares his story. Charlie had a strong passion for teaching and pursued his Ph.D. so that he could become a teacher. As he began teaching, he initially emulated some of his professors who were engaging and entertaining. He quickly transitioned from trying to cover content in his classes to ensuring that students were developing conceptual understandings. He describes using activities to help explain concepts in class, including activities involving candy, demonstration activities, and problem-based learning. He also describes a group of faculty in his department who continue to inspire and motivate him as he continues in his journey to become an exemplary engineering educator and an engineering education researcher.

In Chapter 9, Matthew Fuentes, an Engineering Faculty member at Everett Community College, shares his story. Matthew uses his quirky zeal for learning to create student engagement in his classrooms anchored in a belief in equity and opportunity for all. In recognizing his own privilege in the world as a White, male engineer, he envisions the classroom as a place where all students should be able to see themselves. Through his student-centered approaches, Matthew hopes to change what engineers look like, one student at a time. Matthew's willingness to challenge meritocracy with an appreciation for the process of developing potential positions him as a rare and refreshing advocate for a just education. In finding comfort amid situations of ambiguity, Matthew has enhanced student learning while also cultivating a culture of inclusion that empowers students to reach their fullest potential.

In Chapter 10, we provide a set of lessons learned from the stories. These lessons include taking it slow when innovating in the classroom, finding a community of educators with similar visions and goals, and using reflection to help improve classes. Another take-away from the stories is that innovative teaching can require a lot of work, but can also prove very fulfilling and worth the extra time and effort. One lesson was around focusing on teaching or research, with on story demonstrating that these two aspects of faculty roles can be symbiotic. Other lessons focus around concepts of inclusivity, with one focusing on considering the assets of students when they come into the classroom, valuing their experiences, and being intentional to empower students who have been marginalized in engineering education programs. Moreover, there were many examples in the stories of engineering educators connecting theory through teaching approches to the real world in the classroom through case studies, projects, service learning, and open-ended problems. There were a few examples of engineering educators using concepts from entrepreneurship to improve their classrooms, with a focus on value propositions, considering our customer segments, and pushing on boundaries. Finally, there were many engineering educators who were motivated and inspired to become better teachers because of their experiences as undergraduate or graduate students. The last lesson learned includes a challenge to consider learning something new and trying new things, to help faculty relate better to students in their classrooms who are learning something new and to help expose them to different pedagogies and ways of teaching. As you begin reading these stories, we encourage you to think about lessons or take-aways that can help inform your own teaching journeys.

REFERENCES

Boklage, A., Coley, B., and Kellam, N. (2018). Understanding engineering educators' pedagogical transformations through the hero's journey. *European Journal of Engineering Education*. DOI: 10.1080/03043797.2018.1500999. 3

Campbell, J. (2008). *The Hero with a Thousand Faces*, 3rd ed., Novato, New World Library. 3

Cron, L. (2012). *Wired for Story: The Writer's Guide to Using Brain Science to Hook Readers from the Very First Sentence*, Ten Speed Press. 1

Cruz, J. and Kellam, N. (2017). Restructuring structural narrative analysis using Campbell's monomyth to understand participant narratives. *Narrative Inquiry*, 27(1). DOI: 10.1075/ni.27.1.09cru. 5

Cruz, J. and Kellam, N. (2018). Beginning an engineer's journey: A narrative examination of how, when, and why students choose the engineering major. *Journal of Engineering Education*, 107(4), pp. 556–582. DOI: 10.1002/jee.20234. 5

Meyers, C. and Jones, T.B. (1993). *Promoting Active Learning Strategies for the College Classroom*, Jossey-Bass Inc., Publishers, San Francisco, CA. 1

National Research Council. (2012). *Discipline-Based Education Research: Understanding and Improving Learning in Undergraduate Science and Engineering*, The National Academies Press, Washington, DC. 2

CHAPTER 2

Developing a Liberative Pedagogy in Engineering

Donna Riley

Narrative constructed by Brooke Coley and Nadia Kellam

It's just recognizing that [change] doesn't happen trivially. [It] takes a lot of thought. [It] takes a lot of adjustment. It takes a lot of troubleshooting. And small changes can be tremendously huge… Letting [change] play out organically, it allows for students to shape the class. That's part of it.

Donna Riley is currently the Kamyar Haghighi Head of the School of Engineering Education at Purdue University. At the time of the interview in November of 2016, Donna was Professor and Interim Head of the Department of Engineering Education at Virginia Tech.

CALL TO ADVENTURE: WHY CAN'T ENGINEERING BE TAUGHT THE WAY RELIGION IS TAUGHT?

I think it all started basically in undergrad where I went to Princeton and we had very old school professors there. A lot of them were Oxford and Cambridge educated. They would do the classic thing of taking out notes that were yellowed and 30 years old and write what was in the notes on the chalk board, and we wrote what was on the chalkboard in our notes, and rarely were we ever asked a question in class. The biggest exception to that was a professor who was in a wheelchair and he wrote what was in his notes on a transparency that was projected on a screen rather than on a chalk board, that was the variation. It was extremely passive, and we all had to sort it out later. We learned to work together in groups because all we had was what we wrote down in lecture and we had to figure out how to understand that and make sense of it.

Meanwhile, I took other classes, and this was just kind of my own interest that I thought, well, if I'm only going to have 8 or 10 classes outside of engineering that I'm going to be able to take, I wanted to make them count. I took these upper-level classes in the humanities and social sciences which were over my head, but I just wanted to do that, so I took a class on five romantic poets, so I took a class on women's history in the United States, or something like that. Just because it was interesting to me. I was taking a class, I think it was probably my junior year; no, it was [the] spring of my sophomore year, from Elaine Pagels on Gnosticism and Early Christianity. She had this way of teaching the class where on the Monday we would…She would give her perspective on the reading and we would turn in a reflection paper that was just a couple of pages long. On Wednesday, she would basically facilitate a conversation among all of us in the seminar, there were probably a dozen of us. She would facilitate this conversation among us. What always surprised me was that I felt like I belonged in the room. I felt like I had something important to say. Despite my not having had any of the prerequisites for this class—I didn't speak Greek. I couldn't read things—she'd come in and read, on Monday, she'd be translating from original Aramaic and stuff and everybody would sort of be nodding and I'd be like "How is she even doing this?" Just feeling completely both in over my head, but supported at the same time that I had actually something important to say. Contrast that with engineering [for] which I had all the prerequisites and yet every single time I was in there, they made us feel like we didn't know anything. I became curious [around] that time about why engineering couldn't be taught in the same way that my religion classes were being taught. I didn't really get to pursue that question, it just kind of rested in the back of my mind for a while.

When I got [into] grad school, I found that in chemical engineering at Carnegie Mellon there were people that were much more interested in some pedagogical innovations. They were doing project-based learning and problem-based learning, and they were just more engaged with the literature. A professor named Ed Ko was at Carnegie Mellon at the time, he ended up moving, changing universities later, but at that time he was there, and he was pretty well known in engineering education circles at the time.

It was a campus that was just more engaged with conversations about active learning and so on. I was educated in how to do that. There was a certified program for Ph.D. students from the Center for Teaching and Learning and I went and pursued that. They taught us Bloom's taxonomy. They taught us the basics of what it would take to do active learning. I felt, at the time, I was like "Okay, I can do these things." I was teaching a project-based course that was community-based as well, so we were working with the city of Pittsburgh on Pittsburgh's urban forest. We had seniors in the engineering and public policy program and some Master's students from the policy school working together in teams on how to assess the value of Pittsburgh's urban forest from an environmental perspective. What was it doing to mitigate climate change, having all these trees around? What did it mean for property values? and so on; What did it mean in different neighborhoods? And so on… We were looking at some environmental justice aspects of that problem. I was coaching these teams and really enjoying doing that, and thinking "Gosh, I think I really want to become a teacher." Still not really getting at the heart of what I wanted to understand about the classroom.

It wasn't until I got to Smith College for my first faculty appointment…Smith is a liberal arts college, it's a women's college.… In the fall semester, I taught an intro class, intro to engineering with two other people, so a team-taught class. I stepped in and taught the class the way the other people taught it, but then in the spring I taught thermodynamics. That was really the first time I had my own class, where there wasn't someone else setting the syllabus, the curriculum, whatever. It was just me.

SUPERNATURAL AID: YOU HAVE TO READ TEACHING TO TRANSGRESS

I contacted a friend of mine who I knew through other relationships. This was someone [that] I was an activist with who taught sociology at Grinnell College in Iowa. I said, "Look, I need to understand, like it's time for me to really unpack this. What was different about what Elaine Pagels was doing in my Gnosticism class compared with this other stuff? Because I know something about active learning, I know something about project-based and problem-based learning, but that's not what this is. This is something else that she was doing. What was she doing? What's it called?" I didn't even know the name for it. I couldn't research it on my own because I didn't know what the keywords were. She said "Oh, you have to read bell hooks' book *Teaching to Transgress.* By that point it was spring break, it was March, and I got the book and started reading it. It completely changed how I thought about what was going on in my classroom." It was the key to understanding, not exactly what Elaine Pagels was doing, because I think she might describe her pedagogy differently, but it did talk about the power relationships in a classroom. It talked about viewing students in a holistic way. It talked about valuing the authority of experience and what students bring into a classroom. All of those things were things that were elements that Elaine Pagels was doing in our classroom that were never being done in engineering classes.

BELLY OF THE WHALE: THE START OF A 10-YEAR PERIOD OF EXPERIMENTATION IN THERMODYNAMICS

I noticed that I was repeating some of the very same problematic relationships that existed in my prior experience with thermo. This was true even though I wasn't doing this passive lecture thing. I was doing active learning. I was doing the stuff I was taught to do, but I could tell there were students in front of the class that were engaged, and the students in the back of the class weren't engaged. I could just see it all unfolding in those same ways that I had been taught. The first thing I did was I went back to my class after spring break and said "Here's the problem I've been noticing. I've noticed that some of you are sitting in the back of the class and you don't seem as engaged. I'd like to change the way that we're sitting and so that we can actually face each other. What do you think?" They said "Okay." We started doing that. They felt that was better. I tried this experiment that completely failed which was having them teach each other the material. I said "Oh, well, why don't you just prepare chapter eight and come in and let's talk about it." That didn't work so well, so I abandoned that. It started this 10-year period of experimentation in this thermodynamics class. I was fortunate to be at a place that was, first of all, a brand-new engineering curriculum where we were encouraged to experiment and encouraged to go to the ASEE conference, learn about the state of the art in engineering.

The other really fortunate thing that happened is [that] bell hooks happened to come to Smith College while I was doing this. That was the same semester. There was some connection with the program in Buddhist Studies. She's Buddhist, and they brought her to campus a couple of times over the next few years and one of her visits would happen to be that semester, so because it's a small campus, I was able to go to her. They had a reception for her after her talk. I went to her talk and then went to the reception after. I just walked up to her and said, "Is anybody doing your pedagogy in a science or engineering class?" She said "Well, yes, but nobody has written about it." Then she said, "Whatever you do make sure that you publish what you do." I said "Okay." She didn't really give me any names, so I was able to kind of, I sort of Googled around and tried to find some other folks and I found [a] couple things in science education but nothing in engineering. What's interesting about that is, I wrote up the thermodynamics class, and submitted it to the *Journal of Women and Minorities in Science and Engineering*. It's the only time I've ever had a paper published without any revisions at all.

I fell into this ability to continue to innovate in that class because I applied for a CAREER award to the National Science Foundation. My research area was actually risk-assessment and risk-communication. I was doing technical research in this but my research, because my Ph.D. was in engineering and public policy, was always interdisciplinary.… When I went to meet with the program officers at NSF, I met with the environmental engineering program officer and he said "Well, this isn't really…this sounds more like social behavior and economic sciences, you should go over there." I went over to SBE [Directorate for Social, Behavioral, and Economic Sciences], and the person there was actually someone from my research group, so she had gone on to be a policy school professor in risk communication and was doing a rotation as a program officer at NSF that year. When I met with her, I said "Look, I mean, you know exactly the work I do." She's published in the same area, she knows the area really well. She said "Look, I've got to tell you, don't waste your time. As much as I'd love to fund this, this doesn't fit what SBE funds and I can tell for sure it's not going to fit what engineering does." She said "Don't waste your time writing this proposal. It's not for CAREER," basically. I was upset because I didn't know, you know, what can I do? They told us for tenure, we don't expect to get a CAREER award, but we expect you to apply for them. I was feeling this urgency that I had to submit one but had no idea how or where.

SUPERNATURAL AID: LEARNING ABOUT A CAREER AWARD IN ENGINEERING EDUCATION

Just again by luck, Rich Felder was in the office [of] a colleague of mine. He was in Glenn Ellis' office. I was just walking by, stopped in to talk to Glenn about something else, and Rich, he's just a mentor to everybody, and he took an interest and he said, "How's it going? What are you working on?" and just asked me how it's going. I said, "Well, it's not going so well, frankly, because they told me I have to apply for this CAREER award." Rich has this very famous resource called, "So You Want to Get a CAREER Award," and it tells faculty how to write their CAREER grant. I was like, "I've been told by NSF that my topic doesn't fit." He said "Well, you know you can apply in engineering education." And I said "No, I didn't know that." This was in 2004, and Rich said "Well, you know, they're giving CAREER awards in engineering education now." And, so I wouldn't have known that if he hadn't been there. I had two weeks to write the thing. I wrote it. I submitted it. It wasn't the best grant proposal ever written but they funded it. It was probably a risk for Sue Kemnitzer to do it, but I suddenly had a five-year funded research program that would enable me to explore what it would mean to do bell hooks-like pedagogies in engineering education.

I had used the word liberative to describe these pedagogies because bell hooks did and that turned out to be really interesting. I wasn't quite aware of what that would mean, and it was an interesting move, because apparently that's not a commonly used term. I used it because I wanted to group together various kinds of critical pedagogies—feminist pedagogies, anti-racist pedagogies, pedagogies that are considering class. All of these are grouped together under some label, but probably the term ought to be critical pedagogies. All of that happened. That allowed me to do these little experiments.

ROAD OF TRIALS: BECOMING COMFORTABLE WITH CRITIQUE IN THERMODYNAMICS

The best thing was that grant allowed me to hire a colleague, and so I had this great half time, actually he was quarter time in the beginning, a Research Associate named Lionel Claris. I hired [Lionel] when he was a Master's student in Education at Smith. He had been at Hampshire College before that and did his undergraduate degree on political philosophy, so he knew all of the social theory, and he had an education Master's, and he was now teaching in the K-12 schools in Springfield, Massachusetts.

As we started talking, I was thinking about power relationships in the classroom and the fact that I could ask the students to share power, and they would do it. They would go through the motions of what I asked them to do. I want you to talk to each other more. They would do those things, but they never really seemed to internalize what that was for. Why that was happening? That [it] was about trying to change this fundamental set of assumptions that I knew stuff that they didn't know. That I had some position, the privilege in the classroom, and I was trying to challenge that and mess with that in some way.

They didn't understand that, so he was like "Well, why don't you have them read something about that." I started having them read this piece [that Lionel brought in] from Michel Foucault [1980] on truth and power in science. So, it's specific to science, it lets them think about it in a very concrete way that they can access. It was just three pages. Even if they found it impenetrable, it was short. They would read it and we would unpack it. They took a whole day of class to just talk about that reading and what it meant for the syllabus. What it meant about who decided what was in the thermodynamics class, who decided what was in their textbook, who decided what the discipline of thermodynamics constituted. This ended up being the most fruitful change, and I didn't realize all of the ways it was going to be so important, because at the time I had this focus on pedagogy and I realized with this assignment that in order to change the pedagogy, to really change it, I had to change the curriculum to an extent, and I had to change it at least enough to insert this one reading.

I realized that once I did that, then I had students reading the textbook critically and saying "Well, wait a minute." The textbook that I picked was a textbook that had a lot of real-world examples, so [it was] trying to relate to students. There's this whole unit when they taught the first law of thermodynamics, they taught this piece about energy and exercise and diet, so they'd be like "You're burning calories, you can do an energy balance on your calories in, calories out," kind of thing. But some of the problems that [the book] had them do were problematic from a gender perspective and problematic from a women's health or anybody's health perspective. An example problem was like, "Jack and Jill go to Burger King and Jack orders a Whopper and a large fry and a large Coke, and Jill orders a Whopper Junior and a small fries and a Diet Coke. If Jill weighs this much and Jack weighs that much, how much do they have to exercise to work off their meal?" And Jill has to work way more because she's smaller or whatever. You're learning all these gendered ideas about exercise. Then you've got these other problems where someone diets and loses like 13 pounds in a week, and this one student of mine who was [an] eating disorder survivor wrote a piece. The student was a survivor of Anorexia and she had read this one problem that was about somebody losing 13 pounds in a week and she pointed out that that's really unhealthy weight loss, and so that enabled us to, first, we talked about it as a class, but then that turned into an assignment where I asked the students to pick some of the problems from that section, because a lot of them were really problematic on different grounds, and just talk about them. They could do the problem and then critique what it is saying about health, exercise, whatever, and then write their own new version of that problem, or a wholly different problem if they wanted to, that related to their interest in some other way. That was a really great opening, because the students became comfortable with critique.

REFUSAL OF THE CALL: BECOMING A FEMINIST ACTIVIST

And it led to a second thing where a student came up to me that same semester and said "You know, I read about this thing on the Internet and I don't believe everything I read on the Internet so I just wanted to ask you, have you ever heard of this thing called the Montreal Massacre?" I had heard of the Montreal Massacre because I was a first-year engineering student when [it] happened, and it left a big impression on me, because it was a critical moment where, for me, there was a microaggression associated with the event, so I had read about the event, heard about it. There was a vigil that the women's center on my campus was organizing about [the Montreal Massacre]. I had a chemistry exam that night, so I went to take the chemistry exam, and we were all talking before the exam and this guy in front of me had turned around to see. We were talking and he said "Oh, what's your major?" And I said "Oh, chemical engineering." And he said "Oh, you're an engineer. Where's my gun? Ha ha ha ha." Everybody is laughing. I'm just like did I hear him right? Right. I'm like did he really just say that? Then right then, it was like "Pencils up, take the test." I spent a good amount of time going, "What else could he have said. What the hell?" I walked out of that exam and this vigil was still going on and so I went to the vigil and what was upsetting to me was there was [only] one engineering faculty member there at all, and he was there in his role as, they had these things at Princeton called Masters of residential colleges. There were four or five residential colleges where students live their first two years and they had faculty that had an administrative role at the college. He was in that role, so he was there because he was involved in student affairs. He wasn't just there because he was a concerned engineering faculty member. …. He was the only representative of the entire engineering school. The Society of Women Engineers wasn't there. Nothing. It was all about the Women's Center. That radicalized me in undergrad, and it was a big moment for me because it is pretty much directly how I became a feminist activist. The head of the Women Center found me at that event because I said something [like], "I'm an engineer, this shit just happened to me." She approached me after and said, "You should really come to the Women's Center." And so that started my engagement with the Women's Center at Princeton. Anyway, [back to the conversation with the Smith student] the Montreal Massacre, I was like "Yes, I have heard of this thing." She's like "Well, how come we're at the first women's college engineering program. Like why are we not learning about this event? This is important." Anyway, I said "Oh yeah, sure."

STORIES FROM MY CLASS: THE MONTREAL MASSACRE AS A CASE STUDY

I went in the literature, there was a women's studies class that used a case study, like a memorial to the women from the Montreal Massacre as a way to talk about violence against women. I read [the case study], it was in Feminist Teacher, it's a journal, and so I picked up that journal, read her lesson plan that involved doing a memorial to the women [by] saying their names, and then talk about violence against women. I picked up some important pieces from that, like not talking about the shooter being an essential thing. You don't want to give airtime to that because then you end up in this criminology conversation that you don't want to be having.

   I took the basic structure and some tips from that and then found a bunch of videos from the Canadian broadcasting company that just had the basics of what happened that night and a really poignant survivor retrospective. Because one of the things, the guy literally yelled [was], "You're all women who are going to be engineers. You're all a bunch of fucking feminists. I hate feminists." Right when he shot them. There was a woman who said "No, we're not feminists. We're just trying to get an education." She's pleading with him not to kill them and she was in the original group of ... There's this classroom of about 60 people, and about 50 of them are men, and he orders the men out of the room and there's maybe 10 women left, and he opens fire on them. Most of those women died. She was shot but survived.

They had an interview with her five years later where she's working as an engineer. The woman that's interviewing her is this famous feminist journalist. [The shooter's] suicide note, had this list of women that he wanted to kill that day, but he couldn't access those women. Her name, this journalist's name is on that list. They're having this conversation and [the journalist] goes, "What does he represent to you?" And [the survivor] said "He's just a poor guy." She said, "That's all." She said "Yeah. Yeah. He's just a poor guy." The journalist says "Well, what did you represent to him?" She says "You." She's like "Yeah. You know, like you." And then she starts naming these other Canadian feminists. You, so and so, so and so, but I was easier to take, and then she goes "Well, maybe what we were doing is the same as what you all were fighting for." She's like thinking out loud about, like, "Is being a woman in engineering a feminist act?" She's thinking through this out loud to herself.

It's this incredible moment. It's like a five-minute clip. I played the clip. I'm like "Okay, so what do you think about this?" You get all these different perspectives about people's comfort or lack of comfort with the idea of feminism and they talk about that, and they start talking about their internships. They start talking about intersectional ideas of feminism. They're not just talking about their experiences as women [with] internships. They start talking about how race intersects with that, how class intersects with that, sexuality. They have this incredibly rich conversation that all I had to do was spend maybe 15 minutes [at the] beginning of the class presenting this incident to them, and they didn't get hung up on the violence part of it. They didn't talk about murder. They immediately got the relationship and just started talking about the stuff.

I think it helped that I was in a liberal arts environment because I think they had more tools to talk about this stuff than they might have elsewhere, and this was true of the Foucault reading as well, that they all had heard of Foucault. None of them had read him except one or two that took a sociology class or something. Most of them, they knew their roommates were reading him. They were like "Oh, yeah, I know what this is." There were people they could talk to in their dorms about it. Now they have this opening for conversation outside of class that gave them, I think, a lot of good common ground with other folks to talk about everything and what it means for them to be an engineer and put together their identity with what other people are doing. That's all [a] big aside, those are the kinds of things I was able to do in my classroom that were really a departure from what, at least at that time, anybody else was doing to my knowledge in their classes.

This did empower them to raise questions in other classes. They would come back to me and tell me stories about "Well, so and so teaches really traditionally and I asked him about it." They did learn to push a little harder on some of the other faculty. In the early years of engineering at Smith, I think the other faculty were pretty receptive to that. Everybody was in an experimental place. Everybody was willing to play around with stuff. "Yeah, I'll take your suggestion. Let's shape the assignment this way, whatever, whatever."

ROAD OF TRIALS: PUSHBACK FROM STUDENTS AND COLLEAGUES

As time went on and people got more and more busy with other stuff and less willing or less rewarded for doing stuff with their teaching—whatever it was, they became less interested in that conversation. There was a real shift from this kind of open place to a more like, well, "This isn't how I teach" kind of thing. "This isn't how engineering is."

There was this other pushback that happened where some of the time and in the later years I started to get more and more pushback from the students that what we were doing in that class wasn't engineering...when I asked them to do the Montreal Massacre exercise, when I asked them to think about ethics. The most shocking thing occurred the very last time I taught that class at Smith, [which] was the fall of 2012. [The students] entered a National Academy of Engineering [NAE] competition that was making engineering energy ethics videos. I had the whole class do it. The requirement wasn't that they had to enter the competition, but they had to make the video in teams. That was a semester-long project, [which] was to make a video about some issue of energy ethics, you pick, totally open.

They didn't see how any of that related to thermodynamics. This was despite my spending a lot of time trying to address it explicitly and having those conversations about what [ethics has] to do with thermo. They all entered it. I explained to them what the NAE was, I didn't assume that they knew that. They all entered it, and they won, four teams won awards from the National Academy of Engineering that year. It was a big deal. They won money. They got to go to a conference.

They told the NAE that they didn't think ethics belonged in a thermodynamics class, and then the NAE people got back to me, and they're like "Do you know that they're saying this?" And I'm like "Yeah, I know." It had gotten to the point where there was more pushback. This is a really interesting possible result of the pedagogies I was using.

One of my favorite examples of this happened pretty early on. It was right before Thanksgiving break and everybody is really stressed out. There's a lot of stuff due in all of the classes. She came in and I had handed back a couple different assignments. I was asking them to do learning reflections, which were written, or ethics essays, which were written. And then they always had problem sets, which were shorter because I wanted them to spend less time on them to make room for the reflections. I had this theory, which I would explain to them. It was the rule of the nth problem, a law of diminishing returns. We have you do so many problems that there's a learning curve and as you approach the nth problem, you don't really learn as much when you do it because you already got it by that point. I want them to stop there and then spend their time doing something else. Anyway, I was handing back assignments, and this student got really mad and held up her essay and said, "This isn't thermodynamics." Then she held up the problem set and said, "This is thermodynamics." Or maybe she said engineering, I forget. This is engineering. This is thermodynamics, not this. It was a really poignant moment because that's what I wanted them to do. I want them to be authoritative. I want them to feel like they can decide what belongs in the class, to the point where they're telling me this doesn't belong in this class. That's a good thing. Though maybe the irony of it, I guess, is that they're reasserting a traditional view of engineering in doing that I thought, "Well this is good, I want her to be able to do that."

It was a perfect setup because right after Thanksgiving—and I had the weekend to think about it—right after Thanksgiving, the next topic was co-generation. And they learned it, it's in their textbook, and they're learning how to analyze co-generation power plants. They had taken a tour of the Smith campus plant and we had a heating plant that did not generate any electricity. All it did was, they'd burn fuel oil and then use the steam to heat the whole campus. There had been a lot of talk [about] whether or not they wanted to retrofit that facility to generate electricity and feed some back to the grid. They hadn't done it and there was a widespread belief on campus that they ought to do this and they should have done it yesterday. I was able to talk about it and the students said "Yeah, why don't we have co-generation on this campus, that's so stupid." And I would say "Well, why do you think?" They'd say, "The [Board of] Trustees." So, OK, say I'm on the [Board of] Trustees. What do you need to communicate to convince me?

Suddenly they realized that this was about communication that you had to be able to articulate to someone who's not an engineer, why co-generation was going to save money, why it was going to be more environmental, why it was going to be good PR, why it was the right thing to do. I was able to, the very next class, make the case again for why all of these things were engineering and were really important things for them to know in a thermodynamics class.

I was able to continue to have the conversation with the class, and I liked having that tension because it was a productive tension. There's a type of resistance going on that I was able to make sure that it stayed a learning moment, but then toward the later years, it stopped feeling that way because, I'm not sure. I think some of it was less direct, like the students weren't coming to me directly and saying this in class. They were sort of going sideways, they were telling the NAE, not me, that kind of thing. It became harder to bring stuff back for conversation.

As I said, [I] just got, for whatever reason toward the end, I got more and more pushback. A student told me at one point that one of my colleagues, her advisor, told her that I just didn't like thermodynamics; I didn't like the material, so I taught other stuff, because I didn't like the technical stuff, which wasn't true at all. They said that and that became widespread belief among students so it's hard to refute that one, it's very political in some ways.

ROAD OF TRIALS: REQUIRED SERVICE LEARNING IN THERMODYNAMICS

One of the things that really didn't go over well was the time I did a service learning or community-based learning project where all of [the students] had to go to Springfield, Massachusetts. I thought it was a really cool project. It was about the energy density and energy cost of food, and so we were looking at, we were in a food desert in Springfield. There was a women's studies class that was looking at women's role in organizing this urban farmer's market, where it was a struggle to get the folks from the surrounding farms to drive the extra distance into Springfield. They were filling this critical need for fresh vegetables because there was no grocery store in this neighborhood at all. There were a couple convenience stores, and you just couldn't get fresh vegetables at these places for any reasonable price anyway, and so these women organized the farmer's market there.

My class was looking at doing an energy analysis, because there's this fascinating study that these folks at the University of Washington did—this analysis where on a per calorie basis, vegetables like lettuce cost up to 100 times more per calorie than fats and oils or potato chips. There's this whole explanation of hunger, about why it's cheaper to buy fast food and junk food, and they just have a lot of interesting data. Taking that as the model, can we collect the local data for the farmer's market and compare it to the convenience store that people could get to and then the nearest grocery store which you had to take a bus to, and sort of look at, okay, what are the energy costs of food in this community?

There were some really interesting things that came out of [the energy analysis project] which were that there's an assumption that farmer's market vegetables are more expensive. That wasn't always true and there were places where the vegetables were either cheaper or there were vegetables that they just couldn't get elsewhere. There were a lot of Puerto Rican and other Caribbean families there. They were looking for particular kinds of greens, the farmer's market had those, like to make callaloo. They couldn't get that otherwise. There were a bunch of things the farmer's market was able to provide, and then we presented the results to this larger group of folks, including food bank people, and some of the farmers, and the women that organized this farmer's market.

What was fascinating was the farmers started talking about the policies of the grocery stores in underselling them. Corn comes in season [and] the grocery stores will sell corn at a loss, just to bring people into the grocery store. They started talking about how [we can] counter that. Because the farmer's market is in a food desert, it actually gave them an opportunity, like a market for their corn that they wouldn't have otherwise. There were some really interesting pieces that wouldn't have come out without the analysis that the students provided, but the students had to travel 25 minutes to go to the farmer's market. At the end of the day, the students hated it. They hated it because it was required, I think. Most service learning classes are elective classes, where students sign up for it, knowing they're signing up for it. I said "Okay. Clearly, I can't keep doing that in this class because it's pissing them off." I couldn't justify continuing to do that in a required course. I did service learning in my elective courses, but I didn't do it in thermo anymore after that.

RETURN THRESHOLD: PROTESTING A NUCLEAR POWER PLANT

The pedagogy that I was doing in the early 2000s led directly to my meeting a couple of people who are part of this network on engineering, social justice, and peace. Early on when I was presenting about my thermodynamics stuff, I was in a session with the liberal education division at ASEE with an engineering professor named George Catalano at SUNY Binghamton. He was presenting on peace pedagogy, basically how to teach peace in engineering classes. I was talking about, I think that particular time I was talking about globalization and how to teach critical perspectives on global engineering. We were in the same session and we were like "You! I found a kindred spirit." I invited him to my very first kickoff meeting for the critical pedagogy project, and then he invited me the next year to go to this engineering, social justice, and peace meeting.

I was part of that network of folks and I began thinking more and more about community-based learning and after I did that community-based project with the food bank, I did a different community-based project in my engineering ethics class, which was an elective on the ethics of nuclear power. It was during a critical point where the Vermont Yankee Nuclear Power Plant had just had its license extended by the Nuclear Regulatory Commission. Its life had been extended by 20 years, but there were serious problems with how the plant was being managed, and they had had a number of problems like a cooling tower collapsing. It was rotted wood and rusted bolts that caused that thing to collapse. It should never have happened. I think they were just totally deferring maintenance on that thing. There was a series of ridiculous things that happened. The state came to take an official position of opposition to the nuclear power plant continuing to operate. They didn't renew its public utilities license and said, "You can't continue to operate in the state of Vermont." Well, they challenged this and said "Well, if the Nuclear Regulatory Commission has approved us, we should be able to operate regardless of what the state says." And that went through to federal court and they made a ruling that they could continue to operate.

The state of Vermont actually lost that case, and so then on the day that the state's license expired, there was a massive protest, over a thousand people in the street, and 138 people got arrested, including me. That fall the plant was still operating and the community was interested in continuing to take data. There had been little tritium leaks here and there. Stuff leaking from this plant. People were like, "What's happening when they release steam?" "What's in the river?" "We know it's really hot when they're releasing steam into the river, but what happens downstream, how are fish being affected?" A bunch of questions that were ripe for citizen science projects.

My students started thinking about, okay, "What can we do as engineers to help support what the citizens want to do?" "What questions do they want to ask, how do we do this?" It was fascinating because the students were coming up against differences in what the nuclear industry was saying was valid data and what the citizens (some of whom also possessed expertise in nuclear power) wanted to take as data. They had to confront these different ways of knowing. What did they think was valid, and who do they believe and why? Lots of really critical thinking processes [were] going on and all in this context of citizen science, or as I would call it, citizen engineering.

This is coming up against my own questions about, how do I as a citizen and an engineer oppose this nuclear power plant or get involved in other kinds of projects. My involvement in the engineering, social justice, and peace network led me [to] ask... where are the places that engineers ought to be acting? That was one way that I found that was very concrete and local to me that I could be out there as an engineer and say "Look, I'm a risk analyst. You know, I'm in risk assessment, risk communication. I know about nuclear power risk. You know, people talk about how low the risks of nuclear power are. And after Chernobyl and Fukushima, those numbers were re-calculated. Because they had been based on models, not on experience, but once you factor in all the different accidents that have taken place, the probabilities go up from what the early nuclear reactor safety studies said."

It's not to be alarmist or create undue concern, but the Vermont Yankee plant is a Fukushima clone. It has the same containment problems that the Fukushima plant has, and, no, there's not likely to be a big earthquake, but there are hurricanes, there are floods, and a lot of the same questions apply when you start really looking at the risk and what the evacuation plans are, and so on and so forth. Long story short, I was involved in that and then I went to NSF and the plant closed by the way, they're decommissioning it, which is good.

RETURN THRESHOLD: CHALLENGING THE POWERS THAT BE

I think the cultures in the different places I've worked are more similar than they are different in that there's always a creative tension with traditionalists, however that manifests itself. There's always people who think, "Well this is the way you have to do it." There are unchallenged assumptions everywhere, and I'm sure they exist in my class as well. When you're trying to do something different, you're always pushing against those and you have to push. That is the whole for me, that's the definition of it. If you're not pushing against those, you're probably not being truly innovative. If you're not getting resistance, you probably aren't challenging the right things. You're not being really challenging of the powers that be, if you're not getting pushed back.

That was true at Smith, and it is true at Virginia Tech. You want to be supported. You have to find where your support is, so that you can continue to do that work and I think at Smith, I had that support from day one from the top down because we were doing this new engineering program. I think I didn't have to build it because it was given to me at the beginning. Getting the CAREER award bolstered that support in a big way, so NSF was able to support that work in important ways so that the critics couldn't descend until toward the end of my grant.

At Virginia Tech, there's the same thing of well, we are funded to do this. Stephanie Adams had this huge grant, so sure, I can do it. We're trying to be good citizens to this new roll out of the new general ed curriculum, with backing from the Provost. There are ways in which you go, and you find your support where you can and then go with that. And yeah, if you're doing it right, you're going to hit some pushback. Then depending on who that's from, you address it in different ways, and if it's from students, well, you want to morph and change and do things that are going to be responsive to their concerns like I did with the service learning thing.

If it's from the powers that be, well, [you've] got to really think about, how do you work with them going forward, if it's [that] you're not meeting this requirement or spending too much time here. Well, you always want to just continue to be creative, is the way that I think about it, so there's a new constraint, well, you work creatively with, around, and through that.

ROAD OF TRIALS: CREATIVE SOLUTIONS TO CONSTRAINING POLICIES

One of the big constraints at Smith with doing the collaboration with the women's studies faculty was that if you team-taught, you got half a course credit. We knew that team teaching was going to be twice the work for us because we'd have to somehow join women's studies and engineering intellectually. What we decided to do instead was [say], "Look let's each teach our separate classes, we'll enroll them separately, but we'll have them do joint meetings together, we'll have them do this joint project together." By doing that with a couple different classes, we found a workaround that was successful. We didn't directly challenge this constraint even though it is really prohibitive of collaborative work to say that team-teaching is half the credit. We found a way around it. It's just that attitude of finding the creative solution to stuff. You do want to always be in that give and take of pushing the boundary.

APOTHEOSIS: PUSHING THE BOUNDARIES IN ANY CONTEXT

I don't want to downplay the importance of the institution because I do think I was able to do what I did at Smith in part *because* I was at Smith. I really do think that played a role.

That said, I think there's other ways I would have pushed the boundaries if I had been at a traditional engineering school from the beginning. It would have looked really different but there is this, it does ultimately come from within the individual to do this creative work because it's your class. You're the one that's generating the ideas that are going to push the boundaries.

If I'm advising a new faculty member that's going to one of these places, yeah, you want to go to the institution that's going to let you do the work that you can do without getting in your way too much, but at the end of the day, every institution is going to get in your way somehow. It is about recognizing that and not seeing it in too black-and-white a way. I know a lot of junior faculty are like "Oh, I just have to get tenure. I'm going to just mind my own business, I'm going to get tenure, *then* I'm going to do what I want." But that never works. Look at all the senior people. How did that strategy work out for them? Clearly, this gets drummed out of you if you take that path.

You always have to keep within you the sense that, "Well, okay, I'm going to push the boundaries." Sure, you don't want to do something that's going to get you fired but there's so many things in between doing the one thing that's going to be a bridge too far, and doing all these other things that are going to push boundaries and make people think and challenge the status quo, and maybe make real change as you go.

Or maybe you do want to make that stand that gets you fired. There might be reasons to do that. I would never rule that out, but I think most of the time that's not going to happen. Many people fear that way more than it actually happens, so you do good work, you do what you love. Sometimes, changing institutions ends up being the best route. At Smith, I had a great cadre of folks that I could stir stuff up with, despite the pushback. And, as I engaged more on a national level, I started thinking more about the bigger picture of what was I doing in the field of engineering education, and how could I have an impact outside of my institution. As the Smith experiment wore on, people paid less and less attention to what was going on there, and kept saying "Well, you can do that because you're at Smith. You're a special case." Doing something at Virginia Tech would obviously directly impact a large number of engineers right away and have more influence on the rest of the enterprise of engineering education. That made the move make a lot of sense.

MASTER OF BOTH WORLDS AND FREEDOM TO LIVE: THE IMPORTANCE OF REFLECTION

I think there's something about the reflection of the faculty member that matters in this story. I had a lot of opportunity and still do have a lot of opportunity to talk with others about what I'm doing and why. Having Lionel at Smith and having my other colleagues at Smith too, both inside and outside of engineering, who were able to talk stuff through with me and just be supportive and help troubleshoot and help creatively was essential.

The process of reflection, of really taking the time at the end of the semester, and at Smith this was built into our ABET processes and I still think it's a really worthwhile exercise, and it was built into our grant in NSF too with the science and engineering project. At the end of the term, you stop, and you say well, what worked, what didn't, what's the student feedback, what am I thinking. About how to do this better the next time, what are my goals, how do I think about how I'm going to push the envelope next, that keeps the spirit of innovation alive.

You're not just doing the same thing every year, good, that's done. You're actually taking the time to have a reflective practice about what you're doing and say I need to change this [next] time, more of this, less of that, tweak this, try this new thing here, whatever. It's a constant process. The thermodynamics class over 10 years became almost unrecognizable from the original one, but I never changed more than one or two things a semester. I never completely overhauled the class.

It was about asking just what am I capable of doing and doing well? If I'm going to create a new type of assignment, well, I'm going to do that and take something else out to put it in and just do that and not ... It's just a sustainability thing. You don't want to completely, yeah, you just don't want to make yourself so distraught. I did that community-based learning thing with the food bank, and that's basically the only thing I changed that semester, because it was a big deal to do that.

It's just recognizing that these things don't happen trivially. They take a lot of thought. They take a lot of adjustment. They take a lot of troubleshooting. And small changes can be tremendously huge, like that Foucault change [that] led to all these other changes and all I did was add one assignment. I added one thing. Read these three pages, talk about it in class, write an essay about it.

That led to a whole bunch of other stuff, and so that's being open to that too, and not predetermining. Well, I'm changing this, this, this, this, and this. No, I'm just changing this one thing, let's see what happens. Letting that play out organically, it allows for students to shape the class. That's part of it.

ADDITIONAL RESOURCES

American Society for Engineering Education (ASEE), annual conference. `https://www.asee.org/conferences-and-events/conferences/` 12

Felder, R. (2005). Resources in science and engineering education,. `http://www4.ncsu.edu/unity/lockers/users/f/felder/public/`

Felder, R. (2002). So you want to win a CAREER award. *Chemical Engineering Education*, 36(1), pp. 32–33. `http://www4.ncsu.edu/unity/lockers/users/f/felder/public/Columns/Career-Award.html`

Foucault, M. (1980). Truth and power, Alessandro Fontana and Pasquale Pasquino, interviewers. In *Power/Knowledge: Selected Interviews and Other Writings 1972–1977*, C. Gordon, Ed., pp. 131–133, New York, Pantheons. 13

Hooks, B. (1994). *Teaching to Transgress: Education as the Practice of Freedom*, New York, Routledge.

Riley, D. (2003). Employing liberative pedagogies in engineering education. *Journal of Women and Minorities in Science and Engineering*, 9(2). DOI: 10.1615/jwomenminorscieneng.v9.i2.20.

Riley, D. and Claris, L. (2006). Power/knowledge: Using Foucault to promote critical under-standings of content and pedagogy in engineering thermodynamics. *Proc. of the ASEE Annual Conference*, Chicago. https://peer.asee.org/155

CHAPTER 3

Experiencing Vulnerability and Empowerment in Teaching

Sara Atwood

Narrative constructed by Brooke Coley

Whereas my walking around and coaching them and challenging them and saying, "Now, how does that work?" they didn't perceive that [as teaching]. So, I did get comments, especially from first-years, of, "She didn't teach us anything. I taught myself." Well, yeah. I coached you to learn how to learn. That's the point.

Sara Atwood is an Associate Professor and Chair of Engineering and Physics at Elizabethtown College with specialization in mechanical and biomedical engineering.

CALL TO ADVENTURE: FROM CHILDHOOD TO UNDERGRADUATE, BECOMING AN EDUCATOR FIRST

I think when—you know, hindsight's 20-20, but when I look back at it, I think that I'm kind of an educator first and an engineer second in a lot of ways. Growing up, I never played with dolls, I actually lectured to my stuffed animals and had a little chalkboard easel. My mom was a teacher, second grade, and a lot of people in my family were K-12 educators. So, I grew up around that and always enjoyed school, obviously. In high school, I tutored math for some extra money, and word spread through teachers. I was always really good at math and science, although my favorite class in high school was actually English and Literature. But people told me, "Oh, you're good at math and science, maybe you should consider engineering." I went to Dartmouth for undergrad, and part of why I liked that institution [were the variety of options]. I didn't look at any institutions that were engineering-specific, so I wasn't really thinking about my major or engineering going into college, necessarily. It was kind of in the back of my mind, but I wasn't going for that.

Then, at Dartmouth—a liberal arts school, you didn't declare a major until [your] second year, and at that point, enough people had said, "You're good at math and science, consider engineering," that I took the introductory engineering courses, and I really liked that and really found a home. At Dartmouth, it was a very close—and it's grown a lot now—so at the time that I went, we had maybe 30 students in a cohort, so it was pretty small. I found that it was comfortable, [a] home environment and that's, I think, a big reason why I went in that direction. I feel like, looking back, maybe I could have gone in different directions. But, I liked that. [I] liked the impact that engineering has on the world. And [I] had some professors that were very good mentors to me.

And so there at Dartmouth, I really got that experience of more of the teaching-focused, rather than research-focused. Now, I think it's transitioning a little bit more away from that, which I think is sad. But at the time, the faculty were really teachers first. It was still more lecture-based, but I feel like they were all very aware of doing a good job of educating; it wasn't secondary to them, it was their primary thing. So, I think that they were starting to do [active learning], we would certainly work through a lot of examples and things like that, it wasn't necessarily what I would call active learning now with small group examples. But, it was not just like you only saw the professor's back the entire lecture, and you were just scribbling notes the whole time.

CALL TO ADVENTURE AND SUPERNATURAL AID: EXPERIENCES AS AN UNDERGRADUATE TA

[At Dartmouth] I had a couple of professors who encouraged me, "Hey, you would be a really good professor." And I did some TAing [teaching assisting] there, so I think that might have actually been formative in my ease with embracing active learning, because I ran problem sessions each week. So maybe more like a grad student would do at an institution with more grad students. And in those problem sessions, it was really just small group, walking around [to] people, explaining how to do problems, basically doing that more small-group, problem-solving, active learning model in my problem sessions. And I did grading for them, and had a couple of faculty that kind of brought me under their wing, putting together exams and things like that. So, I got some mentorship there.

So then when I was looking for grad school, I was really specifically looking to go to grad school to become an educator. And I didn't know anything about engineering education programs at that time. I graduated Dartmouth undergrad in 2003, then I stayed for my Master's—so [I] graduated in 2005, then was looking for Ph.D. programs around that time. I think at that time there might have only been Purdue and Virginia Tech, [engineering education] was not a big thing at all and I feel like those were even pretty new, is my sense. I just hadn't heard of that, I hadn't heard of the [American Society for Engineering Education] (ASEE). None of that. So, I was looking specifically at engineering programs in mechanical engineering. Dartmouth was a general engineering degree undergrad, so I had [a] broad base.

So, I thought, "Okay, I'm going to apply to a couple places and be selective, or I'll just go and be an engineer, or even teach high school physics or math, or something like that if I don't get into these grad schools." I applied to UT Austin because I'm originally from Texas, and then Berkeley, because for some reason, I had in my head I wanted to try out the west coast, and I had a professor at Dartmouth who had done a sabbatical there and knew an advisor that he thought would be a really good mentor to me and was doing work I was interested in.

CALL TO ADVENTURE: EXPERIENCING FACULTY WHO PRIORITIZE RESEARCH FIRST

I ended up going to Berkeley, and that was a huge shock because the difference in teaching and student focus from Dartmouth to Berkeley was enormous. My first semester I was like, "Oh my gosh these are the worst teachers I've ever had in my life." And in fact, I feel like there was almost a pride in that [teaching] wasn't their thing. I've known some people who have gone to R1 type places and they've been told that, "If your teaching evaluations are high, you're not doing things right. That's not where you should be focusing your time and effort."

So [I] was just thrown into the deep end of the traditional lecture style. No working examples. You know, the professor's back, working on the boards the whole time. They finish the one [board], they scroll it up, just keep going on to the next one. And I had a really rough time my first semester, transitioning to that. It was very difficult going into professors' office hours and they wouldn't even be there at their posted office hours. And I was like, "What is this?" At Dartmouth, it was an open-door policy. [At Berkeley] you didn't know where they were, you couldn't track them down. I had a really hard time. And honestly, the problem sessions with TAs also were like recitation sessions and not that great either because most of the TAs were not that interested in teaching, either. I think it was a little bit of that rude awakening into that style. It made me, first of all, not want to work at an R1 (very high research activity university) and secondly, really kind of reject that more traditional lecture format, because that was such a rude awakening and it took me a while to adjust to that.

SUPERNATURAL AID: FINDING MY HOME

Then when I was at Berkeley for a longer time, I did some TA-ing [Teaching Assisting] and had courses with a few professors who cared about teaching. I luckily had a mentor (Dr. Lisa Pruitt) who did focus a lot on education, and she sent me to ASEE, even though I didn't have any conference presentation, just to expose me to it because I wanted to go in that direction. I TA'd for her… several times. And through that, I sort of really enjoyed the time that I spent on that, more than on my research. So that was kind of a big clue for me as to what I should do in the future. And, actually, right after I passed my quals [qualifying exams] they had a professor that went on medical leave, and another was going on Sabbatical or something.

[As a result] I got to teach a course at Berkeley all on my own. I was the primary instructor. And I used more of what I had learned at Dartmouth. It was a 115-person class, so it was not the style where you could really do small group/walk around to very easily, and I didn't have, at that time, any exposure for how to do active learning in a larger setting, because I had never instructed in a larger setting. But I did try to do some aspects of it in terms of, my lectures were a lot more on solving example problems, and stopping [for questions]. I didn't know that it was called active learning, then, and I didn't know what the right way to do it was, but stopping and saying, "Okay, now you do the next two steps of this derivation to try to get down to where we only have this and this left," and throwing it back to them a bit. And at the end of that, I got the highest teaching evaluations in the department, and my comments were things like, "Wow, she works examples!" And, "This is like having a discussion section every lecture, it's great! I'm learning so much!" And so that was also pretty formative to me.

It was at the end of my time at Berkeley that I started going to the ASEE sessions and conferences. And I was like, "Oh, this is home, this is heaven, this is amazing!" Every session [I wanted to attend]. So that was huge for me.

FIRST THRESHOLD: FINDING THE RIGHT COLLEGE AND CONNECTING WITH THE STUDENTS

Then I knew, when I was looking at schools [after my Ph.D.], I wanted a teaching focus. I was looking for a liberal arts college, an accredited program, small residential college. There are not many of those, actually. Just kind of a handful. I ended up here at Elizabethtown. And right from the beginning, I was doing a lot more of [the student centered, active learning]. I had been successful doing what I did at Berkeley in terms of just working through more example problems and throwing those [out] to the students in small chunks or whatever. [At Elizabethtown] it was much smaller classes, so that was a little more natural. And I think another thing that really helped was I was closer to my students' age, being a newly graduated grad student. And my advisor had told me that, back at Berkeley, she said, "The years that you are seeming just like an older sister, or someone who could be kind of in their friend group, embrace that. Kind of use that to your advantage." I think that also made it natural to kind of walk around and work on problems with them, and be a little more on their level because we were so much closer in age. And that has changed a little bit over the years, and that's a little bit hard sometimes to deal with, that I'm just getting further away from them socially, so that sometimes makes that gap a little bit harder to close.

SUPERNATURAL AID: LEARNING FROM OTHERS

My advisor had done NETI [National Effective Teaching Institute] with Rich Felder and Rebecca Brent. And she had done that right when I was graduating. I knew of it, and I knew to look out for it, even though our college, Elizabethtown College was not on their list, because we were a newer program and just not all hooked into everything or well known. I knew my first year [at Elizabethtown] to look out for the NETI invitation, and I think I reached out to Rich Felder and made sure that my dean got the invite. And so, myself and another colleague who started at the same time went and did NETI after our first year. And I think that it was actually better after our first year, rather than before you start, because you have a year to sort of see what's natural, see what works, see where you need to pay attention to like, "Oh, that's how I could be tweaking that a little bit." Whereas I think if you get it before you ever do the classroom all yourself, you don't quite know what the more important parts are, the subtler parts, or where you know you have a little bit of trouble and you need to make sure and pay attention to that. But [NETI] was huge. That was very, very formative in my embracing of active learning. But I do think everything that had come before that had sort of primed me to want to do [active learning, and NETI] definitely refined that technique.

Then my second year [at Elizabethtown], we hired another young faculty, and we then sent her to NETI right after her first year because it had been so great. And she came back and she actually embraced the gap notes piece (see Felder and Brent [2015] for more details about gap notes), which I had never done before just because I was doing all new preps. [With gap notes,] what I'm talking about there is basically having a note packet that has blank spaces for where the students fill in, but it provides kind of a structure, a scaffolding, and then there are certain things the students fill in. So, I tend to just type out, if it's a definition or something, I don't want them wasting their time writing out a sentence. Or what's been really helpful to me, I have the objectives right up top, the first page of the gap notes, and then the last page I just have a blank gray box that says "summary." So, at the end of each notes packet, we do a summary and they write it there. And then, a lot of what [the notes] consist of are pictures and problem set-ups. So, instead of them writing down all the given information and the figure, just giving them that information and then spending that time working on the problem. That's sort of what the gap notes are.

At the time that [my colleague] did it—she did it in the fall after she came back and really liked it—and had a lot of success, and I was like, "Ooh, I love this. And now, I'm actually teaching things for the second time, and I feel like I can handle doing that." So, then I really embraced doing gap notes, and that has been really formative in being able to [implement active learning].

BELLY OF THE WHALE: NOT ENOUGH TIME DURING THE LECTURE

One of my challenges, which I think is probably everyone's challenge, is it always seems like there's not enough time during the lecture. So, you're always kind of rushing to get through what you want to and I am a very structured person—disciplined, very ordered and structured. I definitely have an approach of, "I want to get through this amount of stuff on this day because we have weekly homework, and I already have my quizzes set, my exams set, and everything like that." The use of gap notes was really big to me, because it does take more time to let the students work on problems, walk around and talk to them, [and] let them struggle in some places before you then pull them back together and say, "Okay, you tried this out. I see some of you were having trouble with this stuff. Here's how to do that." That's always a challenge.

So, I think before NETI, I was doing a lot of group problem solving, but not in small chunks. It was more, give the class an entire problem and have them work it through, and walk around and talk to them and then come back up to the board and go through it. But, I remember this one student particularly who was very talented, and he always had a paperback book—he liked Clive Cussler. He would get done with the problem so quickly and then he would just open his book. He wasn't being disrespectful, he just [grasped the concepts quickly]. And that was one of the things when I went to NETI that I mentioned that I need to look out for. "How does this work with the timing?" Because at Berkeley and at Dartmouth students were more at the same level, and so people would tend to finish more in the same amount of time. We have a wide range of students' preparation and backgrounds here at Etown, because we don't have a pre-selection program. I was having a lot of trouble with some students would finish a problem really quickly, some would take a long time, and how do I handle that?

With active learning after NETI, one thing the gap notes enabled [was it] put everyone on at least an even playing field to start, because some people just take a really long time to write down the problem and focus and get on it. Some students have learning challenges around writing and processing that make them take longer. And two is that the idea of saying, "Okay now just do this step. Okay let's come back together and go over that. Now do this next step and let's come back together. Okay now finish it up and I'll walk around and help everybody with it." That was kind of a big change. Now it's more back and forth, whereas before it was more like me lecturing a chunk, and then problem solving a chunk, but in a bigger chunk. Now, I guess the engagement is more dispersed. Like me, them, me, them.

So that's become kind of my steady state of where I am now. I've done a little bit of dipping my toe into a flipped classroom kind of thing, so I have recorded some videos. And I try to keep those, I know the literature says about seven minutes is about max for those. I'll try to have two videos each week, which is about a packet or a chapter or whatever, and post those and then just do a 5-minute summary of that, the most important equations, and get right into problem solving.

And this is something that I kind of swing the pendulum on, too. Because I've had students that have responded that they actually like some lecture, and I think one of my strengths is being able to explain things fairly clearly and logically and in a neat, ordered way. So, I've had students that say, "We actually like it when you lecture maybe half an hour and then go to the problem solving," because also a flipped classroom is depending on them to watch the video, which you can't always… And, I think it depends on your style as well. I go a little bit back and forth between, I guess in my lower-level class I do less lecture, more problem-solving because I feel like the concepts are a little easier to grasp quickly. In my upper-level class, I tend to do a little more lecture with problem solving sprinkled in. That's something that I still sort of go back and forth on as well. How much flipping to do and how much outside of class time are students really going to spend with that?

ROAD OF TRIALS: THEORY VS. SOLVING PROBLEMS

One of the things that I still struggle with a bit in terms of active learning—I think that it's best suited with working problems. I swing back and forth like a pendulum semester to semester on the importance of the derivation versus the application in working problems. And so I try to use active learning and say, "Okay, for the next step of this derivation we want to reduce this down to one term, or something," and then try to have them do that for the next 30 seconds or so, and then come in with it and engage them in the derivation. But, there's just a little bit of a debate in my mind about how much [undergraduates] get out of a derivation. I do want them to see where things come from, so it's not just some black box. But, a lot of our engineers also—now we're getting a steady state where we graduate probably 30, 35 a year and two of those will go to grad school—so most of them go out and become practicing engineers. They may be more like construction management, or project managers or things like that. I just kind of struggle with the balance between the theory and the derivation versus the practical and applied problem-solving piece. And active learning, I think, is a little better suited for the problem-solving piece, at least the way that I tend to use it. So that's something that, potentially if I did a redo of NETI, I'd be paying a little more attention to that balance.

ROAD OF TRIALS: JUST-IN-TIME VS. ESTABLISHED PREPARATION

[There have definitely been some trade-offs to this journey]. My first three years was basically all new preps all of the time. Because we're a general engineering program—and when I first got here we were smaller—so my first three or so years, we were teaching classes every other year, a lot of them, like the upper level ones, so each time was new. And, of those courses I was teaching, only one was in my Ph.D. area. So, I was teaching something that I hadn't looked at since undergrad and a Physics 3 course with optics that I had never done in my life. I was doing new labs that I had never done before. So that was like drinking out of a fire hose. It's hard to even say what that was like because it was just so much prep all of the time. I was just a lecture ahead of the students, essentially. But, in some ways that was better, because now I feel like it's a little hard to get motivated to go back and rework the example problems that I've already got prepped. It seems like you shouldn't be spending your time on regenerating new example problems all of the time. I always do new quiz problems and switch out a couple of homework problems. But, when it was just-in-time, and I was learning it almost alongside them, I feel like I was better in some ways because it was a little fresher and I was getting a better fresh perspective on trying to understand the material.

And so, there's been some tradeoffs. I'm teaching things now for the third or fourth time. Things are more settled, so I spend a lot less time on prep and getting to where I'm barely making any changes to [my gap notes]. I'm hoping that I can turn them into a reader. And so that part is nice, but sometimes I feel a little bit disconnected because I'm at the point where now I can kind of go in on Monday and I haven't really looked at the stuff for a year. And I'm familiar enough with it. But then sometimes, it's just a little different than when you're learning it alongside with them two days before. So that's been a bit of a tradeoff, actually. I feel like I've had enough years now that I've been able to chip away at it. I've been able to sort of add something that I wanted to do every year. But I certainly, even though I wanted to do those things, I knew that I couldn't do them all at once. Is there anything still left? I mean the gap notes have taken me several years to get the way that I want them. I'd like to make more videos.

I wish that I could see it more from the students' perspective again. When I was just learning material for a class, it made me closer to their experience. And so that's one thing that I really miss, and I don't know exactly how to get that back. I taught a class I hadn't taught since my very first year, and I changed the format to being mastery-based. It was really energizing to get that new perspective back again, but it was a lot of time. I'm looking forward to tackling that course again next year.

I also think doing more open-ended work would capture that fun again. When I first started, homework solutions were not out there in public. They just weren't accessible like they are now. Students [would] learn a lot doing the problems and would come in and work with me and we would learn a lot through that process. And now, the solution manuals are so readily available online, that only a few students get the same learning out of it. That may be the thing that I would do [to improve as a teacher is] figure out open-ended, interesting, design-analysis problems to do for problem sets, that help them meet the learning objectives, but are really interesting, higher-level struggles for them. I think I might still keep example problems and quizzes simple. But, I feel like the homework sets—in a way—the students who use them correctly, learn a lot from them. But, now with solutions readily available they're set up in a way that students don't have to use them correctly. I've tried doing a couple of "Epic Finales" where students in groups work through an open-ended problem in place of a traditional final exam. It's gone really well.

ROAD OF TRIALS: STUDENTS WITH LEARNING DISABILITIES

I just read "Tomorrow's Professor," [and] it was talking about one of the pitfalls of active learning being those with learning disabilities. It was very interesting. We actually do have a number of students, and I think everyone has increasing students with learning disabilities, partly because of the higher K-12 education system, and our cultural expectation that everyone goes to college. So, [Tomorrow's Professor] mentions though—that these students that have visual or auditory processing issues, or slow processing, or dyslexia—that active learning might not be good for them because it might take a long time for them to process it. So, then, they kind of miss out on the problem-solving piece and it makes them feel worse because their peers are able to do it and they really have no idea what that 15-minute lecture that you just gave was on because they have not gotten time to process it.

Actually, the email was kind of a call for more studies into how active learning might be able to be tweaked for those with learning disabilities or processing speed difficulties or things like that. And that really struck me because we do have a number of those students, and I do wonder if maybe those are some of the ones who are resistant or who don't even make it past the first year of the program because of that way of teaching. I can see [how] psychologically they [could] just feel like, "This isn't for me. I don't get it. Everybody else around me gets it, and I have no idea what she's talking about."

One thing that occurs to me, actually, is [with] the little videos that I do. They are posted and can be accessed any time in the semester on the learning management system. That might be one thing that could help a lot. And I actually have a student who's struggling with English, and that occurred to me, too. "Oh, yeah, the international students, too, who have a hard time grasping the English. They don't have time to translate or really understand what's been said, and now I'm asking them to do something with it like work a problem." So what I've been doing with a student who's been having trouble with English is I actually send him the gap notes a few days ahead so that he has a few days to look at them, to run them through his translation, to look up any terms that he's not familiar with and try to translate it so that he comes into class [better prepared]. That might be a way that students with learning disabilities could come into class more prepared to see it for a second time or a third time and then be able to jump in on the working of a problem better. Especially if those [videos] are closed-captioned or whatever. I can imagine where yeah, that could be some tweaks [to the student-centered teaching] that could have a bit impact.

BELLY OF THE WHALE: A PARTICULARLY CHALLENGING SEMESTER

So many things have changed throughout the years, I guess. I did the Intro to Engineering class for about four years. One semester I had only Intro to Engineering course, and it was just 70 first-years (we expected 50)—all first-years, all of the time—and it was miserable. That was probably my darkest semester because I just did not have much joy in it. And I actually did a workshop with a STEM-UP PA group for academic women in STEM that was funded by the National Science Foundation (NSF) through the Advance Grant. And I remember I was doing this workshop and the facilitator said something like, "What do you take joy in in your job?" And I remember writing down, "encouraging students," and I realized that that semester I was just not doing that, because I had so many discipline problems and [was] on the phone with Learning Services and first-year advisors all the time. And so, for the last about three weeks of that semester after that workshop, I made a point to really reach out and encourage students and write nice emails to them and notes and things. And that kind of got some of the joy back, I guess. But it had just been, like, stamped out of me with that semester.

That was a pretty miserable semester, to be honest. I never want to teach all first semester students again nor just one class with so many sections. I enjoy the variety of topics and student populations. After that the department faculty started sharing the Intro class, which I think was more appropriate—team teaching the first-years.

SUPERNATURAL AID AND MEETING WITH THE ALL KNOWER: A COMMUNITY OF ACADEMIC STEM WOMEN

One of the big supports was the STEM-Up PA group, funded by an advance grant. There was one semester that I did the STEM-UP PA Oasis program. Rutgers [University] has an Oasis program, and it's like an Objective Analysis of Self and Institution Seminar. And so, the STEM-Up program of Central PA teaching-focused group adapted that. And now I think they call it their leadership development program.

It was one semester, and it was four meetings during that semester, Friday nights or Saturdays, for about five hours, so it was pretty intensive. All women, maybe about 25, 30 women. And they also formed peer groups of four women and you also had to meet in person and get dinner or something with those peer groups in between the other sessions. So it was eight sessions during a semester, which was basically every other week. So that was a lot of time commitment, but a lot of support. And a lot of that support, it wasn't necessarily the content of the workshops, it was just hearing other women faculty suffering through some of the same problems and being like, "Oh, it's not just me, I'm not alone." I certainly have experienced it where you return things twice as fast as a male colleague and students have this perception that, "Ugh, she's still grading those?" And you're like, "What about him?" And they're like, "We love him!"

Just sharing some of those experiences was key, I think, to getting me through that. We did a negotiation seminar. It gave me a little bit more empowerment to feel like, okay, next semester, if they try to assign me all of the first-years, I'm going to say, "No, that doesn't work for me. Someone else needs to take some of this on." And not just be like, "Sure! I'll do all of the first-years." And that course content is so much writing and soft skills. The department put me with something that was non-technical, and I didn't appreciate that because it made me look like a kindergarten teacher that was teaching the soft skills, not like someone who was teaching the upper level technical stuff. Historically the class had been taught by the one other female lecturer in the department. After me we started team-teaching with male faculty as well.

MASTER OF BOTH WORLDS AND FREEDOM TO LIVE: A BALANCE OF VULNERABILITY AND EMPOWERMENT

Do I feel like I'm better at teaching now? I feel like my focus has changed a little bit. So that's the other thing about my trajectory. My first few years, the department, the college, the emphasis was really on teaching, so I was really focusing on that. The last few years, then, going up for tenure, I was sort of shifting because my teaching evaluations were very good and very solid. I was willing to accept a little bit of a dip, and some vulnerability in how well I thought I was doing in teaching to get some papers written and get those out the door for tenure. And now,

I'm actually Chair of the department. So now some of my research is really taking off and I've got that going on, and I've got Chair. So, I'm also right now willing to accept a little bit of a lower standard in teaching.

Return of teaching evaluations has really evolved, too. In my first years of teaching, I used to get just so devastated, even though they were really good. But I would get so devastated when there was any sort of negative comment on something. I remember our Dean of Faculty saying, "Well, for tenure, we just want to see that you're reacting to those [evaluations] and you're taking them seriously." And I remember joking to my friend, I was like, "Yeah, let me send you the empty bottle of wine and empty chocolate bar wrapper, that shows you that I'm taking them seriously." My husband was always like, "Oh, did you get your evaluations today?" He knew it would be a couple of rough days. And now, I get them and I read them and I guess I feel a little more evened out about the response to my teaching, and can use them more formatively without negative emotion. I keep a list of positive student comments on my bulletin board for those occasional bad days.

One evolution is in prep work, when I walk in and I feel 80% prepared, I'm okay with, "Well, I'll wing it on the other 20%," that's a little bit of a difference. And I think there's also that realization that the 80-20 rule or whatever, you know, my first few years—I guess I was feeling sort of vulnerable, too, because I was new and the students didn't really know me yet, and I was young, and I got mistaken for a student all the time, and being a woman also walking into the classroom and not having much authority. That's the positive about getting older. I'm like, "Well, I'm going to get gray hair, but I'll get some authority, too." But, yes, I think, just personally, knowing that I didn't know the material that well, I would work extra hard to make sure that I was prepared. I think that I allowed myself to be vulnerable, too, and say, "I'm not sure. Let me get back to you." And the students were pretty good about that. But then, over the years, I felt a little less vulnerable in terms of how good of a teacher I am, or whatever, and so have allowed that whole, "It doesn't have to be perfect. Good enough is good enough." But, it's really satisfying because I still have days where I walk in and I just absolutely kill it on a lecture, and it just goes really well for whatever reason. And I have to tell the students, "It's actually time to leave now, we [have to] go." So those are definitely good days that I still appreciate, but I guess I don't get upset anymore when days aren't like that.

I will say that one of the trade-offs, I think, is some students really like the active learning. But, one thing that strikes me, is that some students are pretty resistant to it. And particularly, when they're first years, you get a lot of the comments of, "They didn't teach us anything," because I think that they have a very specific way of thinking about what teaching looks like. [Students] think teaching is, "Open my brain, pour in your brain" kind of thing, rather than the sort of self-discovery together. So, I've had some students that tell me, "I would rather you just lecture the whole time, and I just want to sit back and take notes." So, there are some that I think remain all four years a bit resistant to that, and would rather be passively lectured at, even though all of the literature shows differently. And I think that might have been my challenge, too, with having first year students, is their perception of what they thought I should be delivering to them as a consumer because education has gotten very consumerized, especially at a private, high-cost institution, [which] was very different than what I felt like my role was in the classroom. And I wonder if, in a more traditional sense, they might have felt like they were getting what they perceived as teaching. Whereas my walking around and coaching them and challenging them and saying, "Now, how does that work?" they didn't perceive that [as teaching]. So, I did get comments, especially from first-years, of, "She didn't teach us anything. I taught myself." Well, yeah. I coached you to learn how to learn. That's the point.

So, I will say that I have seen that resistance. It has not, certainly, been enough to stop me. And I think, when I first got to Etown, part of why I was so successful is because I was teaching some students that had been taught by a visiting instructor that was very lecture-based. So, all of a sudden, when I came in, and now they had that alternative, and I was teaching upperclassmen, they appreciated that a lot. Same thing at Berkeley. I think my evaluations were honestly artificially high there because I was just doing something different than what they saw in all of their other classes. Now that we sort of almost all do it at Etown, I think some of them almost want the alternative where they would rather just sit there and be passive and be given the information.

I think when I look back, I guess the main thing that was formative for me was that experience at Dartmouth, kind of small, with a teaching focus, and then going to Berkeley and seeing the alternative. And I wanted to make that journey, that was intentional on my part, but I didn't realize quite how different the experiences would be. So, I think that that was one of the most formative things in my student-centered active learning teaching style.

REFERENCES

Felder, R. M. and Brent, R. (2015). Random thoughts… handouts with gaps. *Chemical Engineering Education*, 49(4), pp. 239–240. https://www.engr.ncsu.edu/wp-content/uploads/drive/115p3UW6e7oQ_JBOpvMEPcpz8APhxj2Ph/2015-r_HandoutsWithGaps.pdf 31

CHAPTER 4

From the Armed Services to the Classroom

Brad Hyatt

Narrative constructed by Audrey Boklage

It wasn't just something from a book that was 10 years old, but it was something that was currently happening. It was very relevant. Those are the type of opportunities that make me excited, that we can provide students once we make it focused on them and engage technology or bring that technology into class as much as possible.

Brad Hyatt is an Associate Professor of Construction Management at Fresno State University.

THE CALL TO ADVENTURE: A TRUE LEARNER

I've been teaching now full-time for 7 years. Prior to that, I spent about 12 years in the industry, first as a civil engineering officer in the Navy and then a couple years as a project management, construction management consultant, working on large construction projects. When I came and started teaching, I was really interested in bringing my experience to the classroom, and really the only way that I thought I could do that effectively is by talking about projects.

I've always enjoyed teaching. Even when I was in the Navy, I taught some college classes, and then I did some training in which I would instruct other people. I always enjoyed that. I've always enjoyed mentoring others along the way and being a part of that. I think a lot of it has to do with just my makeup, and the way that I am, and I enjoy that. I also enjoy learning new things. I think any of us that are in academia, that's a big part of why we do what we do, is that we really, we have those questions, and we want those questions answered. We're really good at learning. A lot of us wouldn't be in this position if we weren't. For me, the process has been to really try new things, and always having that goal of "How can I improve?" "How can I be better?" "How can I take the feedback from students and from my peers?" "How can I look at the examples of what else is going on and try that?"

One of my personal goals had always been to teach at the college level or a university level. I'll be honest, I never thought it would happen this soon. I had envisioned it would be something that I would be a consultant for 10, 15, 20 years, and then go back and teach 1 or 2 classes as an adjunct.

The opportunity came 7 years ago. This position came open. My background is the Navy. As a civil engineer in the Navy, they send you to graduate school, so I went to the University of Texas, got my Master's degree in construction engineering and project management. At the time, here at Fresno State for this position with construction management, they were looking for someone that had industry experience and a Master's degree as a minimum. It was just one of those things that the door opened, I went and interviewed with it, it worked out, and they offered me the position.

REFUSAL OF THE CALL: DECIDING TO LEAVE INDUSTRY

Quite honestly, it was a very, very hard decision, because I loved my job. I loved the company I worked for. I loved the project that I was assigned to. It was a very exciting project, probably one of the best projects that I had ever been a part of. It was a hospital project in Riverside, California, just a great team. There was about a year and a half left in the project. I just looked at really what I wanted to do.

As far as my professional experience goes, specifically with the Navy, as an officer in the Navy, I think what that has allowed me to do, it's given me the confidence to try new things. It's allowed me to understand and know that I can try new things. I went and talked to my boss. He said, "Really you should take it. It's a unique opportunity that may not come along in the future, at least the way it's designed." I took his advice, and it's been the best decision that, one of the best decisions I've ever made in my career.

ROAD OF TRAILS: CONNECTING CLASSROOM TO INDUSTRY

I would say that, when I first started [the faculty position], it was a huge transition, not just for me but for the students as a whole. I got lots of negative feedback from students, comments, you know, "I'm not paying to work in class. I'm paying you to lecture and teach me things. You're not like other faculty members. Your class is too hard. I don't like the style where I have to do homework or problems in class. I'd rather do it on my own time." There were a lot of comments that I got back initially that were discouraging. What I found after my first semester of teaching was that I just didn't like lecturing. I didn't like being the person in the front that talked 90% of the time, and students weren't paying attention or asleep or just not engaged overall in the classroom. It was very frustrating.

I decided quickly that I needed to do something different. I needed to engage the students more. I needed to get them more excited about my profession and what they were going to do eventually. At that point, I started to do some research, reach out to people here on campus and other places to see what they did. What I quickly found was that there were other ways, primarily project-based learning and things like that, but the feedback that I kept on getting [from other professors] was that it's just a lot more challenging to do that.

CROSSING THE FIRST THRESHOLD: FLIPPING THE CLASSROOM

Being a new faculty member, and with lots of vigor and excitement, I decided, "You know what, I'm going to go ahead and do it anyway, and see how it goes." The next semester, the second semester that I was here, I started to think about and utilize project-based learning. You bring in projects into the class and have the students work on specific projects and do more hands-on work, and a lot less lecturing. Then the next year, I found out about flipped teaching. I decided I really liked that approach where the students would still get the content, but it would be delivered outside of class. Then when they came into class, we would do things with that content, whether it be work on problems or work on a project itself. I really started to see my students get the content and get the problems and do much better on the exams than what they had previously done. Ever since then, that's really the approach that I've taken.

APOTHEOSIS/FREEDOM TO LIVE: LEARNING TOGETHER

As a civil engineer in the Navy and also a project consultant, that has given me a very broad-based experience. When I go into the classroom and I talk about certain things, I just have a very broad background in which I can pull from, not that I know everything, because I don't. That's the very first thing that I tell students is I don't know everything. If I don't know the answer, then we're going to find the answer together. Luckily with technology, computers, and Google, you can get an answer to almost any question. It may not be the right answer. You may have to dig for the right answer, but we can get an answer and really try to discern what the correct solution is or answer that question. Again, having that broad-based background gives me the confidence to step into a classroom and know that either any question that a student asks, either I'm going to be able to pull it from my background or I'm going to know the resource that we can use to answer that question. It may not happen in the class, but soon after class, we can get that solution or answer.

SUPERNATURAL AID: PROFESSIONAL KNOWLEDGE

If I hadn't had that professional experience, if I hadn't been in the Navy for over 9 years, and then been a project consultant for a couple years after that, I don't think I would have had [professional knowledge]. I wouldn't be able to do those things as confidently as I can do now. It's really a culmination of all my experience and just a willingness to try.

One of the things, especially in our discipline, in construction management, in construction engineering, and management, that we always tell students is that our discipline is one in which you work with people all the time. We do not have the opportunity to work in a silo. We don't get to work by ourselves in a cubicle without ever having any human interaction. That is not the reality of our career path. I try to explain to them that the classroom itself, I am doing my best not just to deliver the content, have them learn the content, and learn some of that technical and management aspects of what their job's going to be, but also I want them to learn those soft skills of how to interrelate, how to work in teams, how to articulate your position and work with others. Really having this focus on student interaction, group work, project-based learning, it provides that opportunity.

MASTER OF TWO WORLDS/RETURN THRESHOLD: REAL-WORLD EXAMPLES

What's been most rewarding in this process has been not necessarily what I had planned students to get out of it, but when we go beyond what was planned in the class, and we really start to talk about or do work problems or look at projects that have much more depth than what was originally planned. A prime example of this would be when we talk about construction law in our classes. Sometimes law can be a very tricky subject, especially for a student that's fairly early on in their curriculum. We do an introductory law class, construction law class, at the sophomore level. They're very challenging topics a lot of times, but by giving students the tablet, what we allow them to do is do some research. When we have a specific topic in law, have them go and do a web search and find some examples, and then we can talk about those examples. The examples then are contemporary examples. I can think of one example where it was a case that had come out within the last month, and we can really talk about what that case was, and how it applied to our topic, and it became something very real that the students understood. It wasn't just something from a book that was 10 years old, but it was something that was currently happening. It was very relevant. Those are the type of opportunities that make me excited, that we can provide students once we make it focused on them and engage technology or bring that technology into class as much as possible.

FREEDOM TO LIVE/ULTIMATE BOONE: CONSTRUCTIVE CRITICISM

Do I still get negative comments [about my teaching]? Absolutely. I always do. There's always some. There's always 1 or 2 that just don't ... that really don't like that style, especially the students that are used to learning on their own and used to not having to work in groups. Really, they like to process things by themselves. What I see with those students, and it could be students that are really good, and it could be students that aren't as adept academically, but I always get some feedback from them. "Why do I have to do this? I can do this work on my own. Once I get the problem right, why do I have to help someone else? I have other things to do."

Maybe I'm unrealistic in thinking that I'm never going to have a semester where I don't have some kind of comment that really critiques what I'm doing. Come to think of it, really, one of the things that I always tell students is I want your feedback. I want your critical feedback on my class, on the content, and on myself. If you think that there are things that I can do better, I want to hear it. I welcome it. I feel confident that I can try new things, that I can do something different. That's reassuring. I'm not afraid to fail. I'm not afraid to try something new.

SUPERNATURAL AID: FACULTY SUPPORT

The past two years really has been the most exciting, because here at Fresno State, our new president, President Castro, had an initiative to bring tablets into the classroom. The way that the university approached that is to have faculty really decide how they were going to do that, and they provided us with the training and the resources to explore new ways to bring tablets, technology, into the class, engage students more, and hopefully improve the learning atmosphere. This is the third semester in which I've done that. It's been extremely exciting.

What Fresno State has done is they have created programs in which they provide all kinds of workshops and opportunities for you to become a better teacher. Like a lot of institutions, they provide these faculty learning committees or faculty learning groups that focus on a specific topic. I've been involved in a number of those, one on flipped teaching, one on using e-portfolios.

Then they also have a lot of programs that allow faculty to redesign courses, to integrate technology or new teaching practices. Those oftentimes are a full-year program. You meet about once a month for a couple hours each meeting, and then you go to a "summer institute," which is a week-long intensive program in which you do the heavy lifting and redesign your course. It's a really good program that Fresno State has created. Even if you come in and don't have a lot of teaching experience, they provide a lot of resources to help you get better at it.

ROAD OF TRIALS: RESISTING CHANGE

Again, there's a lot of bumps in the road. There's a lot of frustration. There are some students that just don't like [changes]. What I've seen this semester is that almost every single student engages it. They're excited to have a tablet. They're excited to have a tablet, because when I do problems, when I do activities in class, they have something in their hands that allows them to leverage what I'm talking about and do more than just what they could do with a piece of paper and a pen. Really, I view it as the next step of having technology in the hands of all students and engaging them more, and really moving away from what I'm doing in the class more to what I want students to do and what they can do.

FREEDOM TO LIVE: EMBRACING THE CHANGE

I just truly believe it's so important to have the confidence to try new things. A faculty colleague of mine, he and I were talking about it yesterday in that it's so much easier to do lecture. It really is. In the big scheme of things, if I was solely focused just on myself and not really…not that I wouldn't care about the students, but if I was less focused on the students and really focused on me and my time and being efficient, I would lecture every single period. I would lecture and give quizzes and give an exam. It's just so much easier that way, but I don't learn that way well myself, and I find that I have to be engaged and I have to be interested.

The work load is immensely higher than traditional teaching, but I think, just from my standpoint, I really see huge benefits to the students. The feedback I get from the majority of the students is, "Wow, this is great. It's one of my favorite classes. You're a great instructor. I appreciate you." Then to see them in a next class down the line really understand the concepts and be able to apply the concepts, that just makes such a huge difference. I think that, by trying something new, using technology, using project-based learning, doing some of these things that are innovative and out there, it takes a lot more time and energy.

MASTER OF TWO WORLDS: INVESTING TIME

I would say, for anyone, if they try something new, really put the time and energy, know that it's going to take a lot of time and energy. Any time you try something new, outside the box, it's going to take a lot more time, more time than you probably anticipate, but if you stick with it, if you get feedback from students, if you have them involved in the overall process, it really does pay dividends.

FREEDOM TO LIVE: GAMIFICATION IN THE CLASSROOM

My next step is I'm really interested in gamification, and finding a way to integrate points, [badges] into our courses. I tried a little bit of it in the past, and it's challenging. It adds a little bit more complexity to the classes, but I'm really interested in finding a way to streamline that. Hopefully that will take the students further and get them more excited about it, the course itself, and the content that's created and the projects that come out of it.

CHAPTER 5

Engaging Students through Service Learning and Innovation

Chris Swan

Narrative constructed by Brooke Coley

To me, a faculty member is someone who actually provides the best education that they can for their students. So, doing it where people are engaged is a key part, at least to me... What I really think is engaging the students [is] doing it in such a way that allows them to take the technical expertise that they've mastered in the classroom and apply it to real-world situations and learning as well. Not just doing the technical expertise, but actually learning about both their technical and non-technical professional skills. That's the best education I think we can provide to our students.

Chris Swan is Dean of Undergraduate Education for the School of Engineering and an Associate Professor in Civil and Environmental Engineering at Tufts University.

CALL TO ADVENTURE: THE TEN-YEAR PLAN TO BECOME A PROFESSOR WITH PRACTICAL EXPERIENCE

I started off at UT-Austin. I grew up in Texas and always dreamed to go to college, period. I chose engineering because I found it to be closest to my own interests and capabilities. I always enjoyed math and science, and since my father was an excavating contractor, I said "let me do something in the construction field," and civil engineering fit best as a direction. So, I pursued that as my degree. Struggled through it as a lot of people do, but by the time I was a junior, I finally got the hang of all this stuff because it became more applied. And that was the key for me, was that all of a sudden, I started to see the application of all that math, science, and other often abstract things that they require you to take.

[I] finished with a Bachelor's, went straight through for a Master's with the particular disciplinary aspects of geotechnical engineering. My plan was to then work for a while and return for a doctorate and then an academic career. [I] graduated in 1986, turned in my Master's thesis, loaded up my car and drove to Massachusetts from Texas. I arrived here to work for a company. Originally, [it] was going to be for 4 years, but it turned out to be only 3. Then, went back to school (MIT) to get my doctorate so that I could actually go on to the academic profession.

After MIT, and 5 years of doctorate education, I was lucky enough to get a position here at Tufts. Amazingly, I did have a 10-year plan after my Bachelor's to actually become a professor and it happened. It was just a number I threw out at the end of my senior year saying, "Yeah, I want to pursue a Ph.D., but I want to pursue a Ph.D. with practical experience because that's what's helped me to learn things." So, I wanted to go out and work for a little bit and take the knowledge that I gained from working…It [was] 3 years of practical "apprenticeship," knowledge that I could bring back into engineering education.

CROSSING THE THRESHOLD: HELPING STUDENTS CONNECT THE THEORETICAL AND PRACTICAL

I've [now] been teaching here at Tufts for [more than] 24 years. What and how I began teaching were basically the same way that I was taught. But I always had valued, at least in my own learning, the application aspect. So, it wasn't just "Here's the formula" and trying to get to the mathematics of the formula, but saying, "Here's the formula, let me tell you why it works and how it works and where it is applied." And then you can work from the "I've applied it" to "Oh, let me try and understand the mathematics of it, or even the science of it," so it was a reversed direction. What that has shown me is that the math and the science is so important to understanding many of our engineering principles, but engineering is still something that students need to also experience—to the point that the application is so important that it makes the math and the science that much more interesting to them. In other words, I can actually apply that calculus to that, or the differential equations to that. But they also get to see this particular principle, if you will, and its real-world aspects and its applied aspects.

The fact to me is that connecting the theoretical and practical aspects is important. It's not necessary, I don't think, for all students to experience engineering as such a connection, but for me, it made the experience that much stronger. [Students think], "Oh I just learned this formula, and I can easily plug and chug in this formula, but I don't understand what that means in reality," so, [I] try to bring in real situations where that can happen. And the real situation could be that you do a video, or something that they actually have to do themselves. To me, the tactile [nature] of actually making something makes a difference in how students will engage with a particular topic.

As a civil and environmental engineer, let's say [the lecture topic] is water resource related, I can't have them build an actual dam. [Nor can I] really have them do watershed analysis from the standpoint of delineating its location for a particular river or stream. That's a good field effort, but it's not what I'm doing. And I would say [in] environmental studies people can do that, that actually fits very well. But for me, I can show it in a planned situation, and then talk about the situations as they arise in reality. Not just, "here's a plan," but we also can see that, here we are in Massachusetts, close to the Aberjona River watershed. So, what does that mean? Well, the Aberjona River watershed is the same one that is in the book (and later movie) Civil Action from the 1980s/late 90s. But it has real-world implications because understanding the Aberjona River watershed was an important aspect of this case of contaminating a community's drinking water, leading to an increase of cases of leukemia. But it makes a connection because this is real, the people are less than 20 miles from Tufts, and that impact is real. So, if I can make those connections between the abstract concept of watershed analysis and the concrete reality of understanding a watershed so that you can see its impacts on the community, I think that it just brings home the topic. It makes it deeper for a student to go into how they can understand the subject.

APOTHEOSIS: SEEING AN EXPLOSION IN THE DESIRE OF THE STUDENTS TO LEARN

So, my teaching was not completely hands-on, but I did a lot of projects at that point to make it hands-on. But it was still, and I would say, it still is, strongly lecture-based. But, the evolution of my lectures is another piece that's interesting ...I would say that the real evolution in my teaching, and almost revolution in my teaching, occurred when the projects started to become real projects instead of ones that I had made up and controlled the data.

In [the] spring of 1999, I was still teaching a course called Site Remediation Techniques, which are basically methods in which we clean up hazardous waste sites or toxic waste sites. Previously, I had always used projects for which I knew the result; either the site had been cleaned up and I had the data, or I had made up the data to lead to a specific solution. For example, here's the soil profile, here's the chemicals of concern, let's think of a method [with] which you can clean it. How would you go about cleaning up this thing? The "switch" in 1999 was that we were working on real sites with no known or given solutions. Additionally, remediation of these sites would have impact on the economies, the social fabric, all different aspects that weren't traditionally seen, nor taught as an engineering aspect, beyond just technical [aspects]. So, these projects were basically service learning-based projects. We were doing projects in service to a community, working with clients, working with regulators, working with the community. And so, when I introduced those as projects, I saw the change more so in the students [with] what they learned than anything else. So, all of a sudden, the students became extremely attached to the project. It wasn't just because it was a real-world aspect, but because it had real people.

Students just seemed to love it. They loved it to the point where they were learning things beyond what I taught them in class. Instead of saying, "Oh, we learned this in class, we'll do this method of analysis," it's like "Well, we did talk about that, but that's not what you need to do here." They would actually think about how to implement solutions that we had not talked about. And therefore, they had to get the details I did not talk about—excavation support systems, or methods of remediation, or decision-making processes; items that we had not even discussed in class.

Seeing this explosion in the desire of students to learn is what first got me interested in the pedagogy of service learning. For a number of years, it was just trying to orchestrate the class so that all planned subjects could be completed. Now, it became the logistics for finding potential sites with potential clients and stakeholders and getting students to interact with them. I will say that one of the most powerful achievements of these service-based projects was in the Spring 2000 term, where we had a group of students who not only took a hold of the site, but they actually became advocates for the community. They would go to community meetings and act as technical advocates for the community, and in some cases, get into arguments with the contractors about what should be done and shouldn't be done. They basically provided a tremendous service, in that case. And it just so happened that those students were also graduate students, they were all Master's level students, many of them had worked for environmental agencies or environmental groups in the past, and so they already had a passion for this, and now they had the technical knowledge to go with their eagerness, passion, and desire for social justice and fairness; [they were] a very good group.

Once I started seeing the students who were performing at an enhanced level technically, as well as otherwise, I started to ask the question: Why does this approach—having these service learning-based projects—really engage these students? And, therefore, it went from the way that I taught, to the research direction.

ROAD OF TRIALS: RESEARCHING NEW WAYS TO ENGAGE AND DEEPEN LEARNING

The thing that I've been working on for the last few years is how to make sure that I am not only achieving technical learning outcomes through service-based efforts, but that this achievement is equal in level as found using traditional, non-service-based pedagogical approaches. So, now I'm doing research on the impacts of service on engineering students. How that sort of engagement can actually, hopefully, lead to a better prepared engineer, both technically as well as all other aspects. I call [these impacts] professional skills, other people call them soft skills, but I tend to say you communicate better because you know who your client is [and] you can actually communicate with them as opposed to just talking at them. You will take into consideration things such as social issues and economic issues and political issues, and not just say that that's someone else's job. In doing the research, it became clear that there's more than just service learning that can engage a student. And so now I'm finding ways to engage students throughout the course, instead of just saying, "Oh, you're going to have this really cool project, just wait. 8 weeks of lecture stuff, and it'll pay off, believe me, you'll get to see it."

For example, I now look for ways of engaging and deepening the learning without the effort being service-based nor a long-term project. It's engaging them in the moment; within a class period or about a particular topic. For example, you say to the class "let's design a flagpole," so the class will do that design very quickly; back-of-the-envelope style, using quick and simple calculations, by assuming the flagpole is a simple cantilever beam. You do the calculation, and now you're done—technically. Let's think deeper about this. Is that flagpole supposed to be there? What's the flagpole for? We're now getting to questions not of the design's technical aspects—how big should it be, what type of material should it be, etc.—but, into the why it should be. Is the client really looking for a flagpole, or are they really looking for something else? Would the neighborhood accept that particular location for a flagpole? Does the town have the money to pay for said flagpole? So, other issues start to be seen. And I don't dwell on them, I don't make them the entire discussion, I just make them a part of the topic of designing the flagpole.

A specific case that I used to do was a bridge. I had a project in a sophomore engineering course for the students to design a bridge. But they're not designing the entire bridge, just designing an element; [a] straightforward, simply supported beam. I'd push on the technical analysis by asking them to do things beyond what they know. "It's reinforced concrete. We're not going to talk about it, you're going to have to figure it out." The project then asked for students to create a miniature model of their design, using concrete and created formwork. This is now getting closer to real-world implementation. But then I add other considerations that are not strictly technical. Questions such as, how do we make this sustainable? Should we consider this with the neighborhood or the community's input as to what's necessary? The goal is to get students to recognize that such questions should be a part of the design.

So, when they get into their senior year and do their senior capstone design, if they have to do a bridge, or any structure, they will start to ask those questions. Hopefully, students realized that it comes down to, "what does the client want?" [These are] the first questions that should be asked [along with] why do they want it? How does that structure "fit" with the client, the neighborhood, the bankers, whomever else is involved?

ULTIMATE BOON: BECOMING THE BEST FACULTY MEMBER THROUGH STUDENT ENGAGEMENT AND INNOVATION

I think what [students] did recognize was that the remediation course and the course's material became much more interesting and connected for them because the class changed. It increased in number, essentially doubling in size once the service-based efforts were integrated into it. Additionally, its audience changed. When I first started teaching, it was predominately graduate students who did not have strong technical backgrounds. Then, it switched over to be undergraduate seniors who had stronger engineering background, but little practical experience. Then when I started doing these real, service-based projects, it became more balanced; about half undergraduate and half graduate students. What I was seeing was graduate students coming into the course because it had a stronger connection to actual case studies for them. That is, at this time, many of our graduate students were part-time [students] holding full-time jobs.

So, the course was taught in the late afternoon/evening, allowing them to be involved. Many of them were working for environmental agencies and consultants, but they saw that this was a good course that would help to hone in some of the things they had seen out in practice. What I found really interesting was they were getting technical aspects from me, but they were bringing to the classroom their real-world experiences. So, in essence they were co-teachers.

[Graduate students] could talk about, "I remember this site, we actually pumped it, and we actually found this and this ..." Yes, real world. They may not have understood the theoretical or technical aspects of pumping; that is, as you pump and you get a draw down and you can actually figure out the change in height of the water at different points, that technical detail [was] not there. But then they could make the connection of, "Okay, when we pumped, we saw the water level being different across the site." They could make that connection to what a technical analysis was saying. So, I think it created an opportunity for students who had real-world experience to make those connections. It also created, especially in this mixed classroom that I had, an opportunity for more 'default' instructors to be there, and to be educators to those whom [had not yet had] such real-life experiences. After doing it for 2 or 3 years, [I could] see the value. These people [were] not just picking up a real-world project, they were actually providing their expertise and their knowledge in a non-technical sense to what the technical issue was. So, the undergraduates that could do the technical work, calculations out [of] the kazoo, but they didn't understand what the calculations were for. Whereas these graduate students, especially the experienced ones, they could and they could run with it.

The impact on me and my time was substantial. Because, to do that, to maintain that same level of technical competence, and to expose these other things, was additional work—additional work on the faculty, additional work on the TA, additional work on the students. But the students won't see it as work, if they see it as learning. So, there are barriers; I call them self-imposed. It depends on your own personal value proposition. I don't have a personal value proposition that says I need to become a full professor and then write papers all the time and have a graduate [cohort] of 10–15 students. That's not my personal value proposition. Mine is delivering an education to all students. And this allows me to do that. Yes, it takes a lot of time. More so than what some people say I should be doing, probably. I agree with that. So, the barrier in my case has been my own personal goals and interests. They are internal as opposed to external. Intrinsic versus extrinsic.

To me, a faculty member is someone who actually provides the best education that they can for their students. So, doing it where people are engaged is a key part, at least to me. Do I find colleagues that pursue this as well? Yes, and more and more of them are coming. But it's not an overnight sensation and everybody wants to do it, no. Not that way. What I really think is, [engage] the students in such a way that allows them to take the technical expertise that they've mastered in the classroom and apply it to real-world situations and learning, as well. Not just doing the technical expertise, but actually learning about both their technical and non-technical professional skills. That's the best education I think we can provide to our students.

So, that's where I am right now, I have academic evidence to show that [service learning] works. [This thing I did in 1999] is still impacting [my ability] to continue on these different pathways. [I'm] still working on things, got an entrepreneurial side to it, too. Most people look at entrepreneurship as another way to say, "I want to make a lot of money and I want to make a lot of money fast." To me, entrepreneurship is finding out what your client wants—basically their values—and saying how do I satisfy those values? And you may find out that what is currently available doesn't satisfy them. So, you have to be innovative in the process. Entrepreneurship is truly a mindset where you really evaluate if something can be grown, scaled and sustained. Why not be entrepreneurial in applying an educational concept? An educational innovation? When most of the education that we still receive today is the traditional lecture style, when people can deliver it in a different way, why can't that be an entrepreneurial effort? My value proposition is that service-based efforts enhance student's learning outcomes. I'm looking at service learning as something that engages them so much, and they continue to be engaged by it throughout their lives, that they say, "I picked that up at Tufts." Same thing at another institution, "I picked that up, at Institution X." Long term, [service learning] could have broad and deep benefits—[and] this is really long-term thinking—[as] an engaged student leads to an engaged alum, which leads to [a] continued flow of institutional support. [Such efforts will influence a] different student body, hopefully more engaged with learning, but also more engaged with the institution. I want to say that it really comes down to wanting to deliver the education that I think is appropriate and most impactful to the students. What I'm seeing just by doing it and being involved in it, is that [service learning] is impactful.

CHAPTER 6

From Food to Simulation with Legos: Engaging Students in Hands-On Learning

Thais Alves

Narrative constructed by Audrey Boklage

Right now, I hope I can better manage this struggle that I can just smoothly put these innovations in the class and still be able to move on with the content. This was another thing that the CTL (Center for Teaching and Learning) changed in my mind before I wanted to go and bang, bang, bang, bang, cover the syllabus. Right now, I said, "You know what? If they learn this and this and this and they really know well about this, I can skip a topic or two and maybe have a smaller amount of time dedicated to that." I think that my battle is to change a little bit at a time but still cover the material.

Thais Alves is an Associate Professor of Construction Engineering at San Diego State University.

CALL TO ADVENTURE: CREATING A COMMUNITY

It started all the way when I did my Master's in [lean construction] in Brazil. Then I went to UC Berkeley and I had more exposure to [teaching]. Through my time there I also TA'd for my advisor and I really enjoyed being a TA and working with the students and seeing how the assignments were prepared from the other end.

After my Ph.D. studies, I had to go to Brazil because I had a scholarship, a full ride. I had to go there [Brazil] and spend some time. Then when this opening showed up at San Diego (San Diego State University, SDSU), it was in an environment very similar to the one I had in Brazil in the sense that the industry there is extremely supportive of our program. Whatever we need, if we structure the call or the request nicely, we'll get help. Projects that we send people to collect data, research, whatever, you name it. It was the same thing when I was in Brazil. The interesting thing when I was [in Brazil, there] was this group that was very close to the university down there, they had their own learning community, so there were 10, 12 companies that paid money every month to become a member of an innovation type of community, and they would bring experts from other parts of the world to see what they were doing, and they were very creative in how they implemented lean construction, so much so that people from around the world, we were often hosting people just to show what they are doing. One of the things that caught my attention very early was that people who were trying to explain lean [construction] to us, they would always have different ways of explaining. It was not your traditional "I teach, you sit and learn." There were a lot of fun exercises outside of the classroom and games and the sheer volume of discussion and how we were trying to understand it, because we are engineers and we were trying to understand all this philosophy behind lean.

I think the ingenuity that they have in the U.S., everything has to have a computer and a laptop and a tablet and a projector. [In Brazil], this is not [the case]. People [become] very creative in terms of how they [do] things with paper and pencil and conversations because of [the lack of a computer and other technology]. That was very good, too, because when I was there, I had a bunch of examples that I could give to my students and they could step out of the university and see it.

ROAD OF TRIALS: A LACK OF IN-SITU EXAMPLES

When I came here to San Diego in 2009, I was talking about some of the concepts that I taught in my graduate course [in Brazil], and people had never heard about them. I didn't even have construction sites to send [students] to see because [they were] not here. That forced me to be even more creative on how these things were going to be put together because they couldn't step out of campus and say, "Oh, we are going to go to a construction site and see these." Some of the concepts that I teach them, to this day, they haven't seen anywhere. My students back in Brazil, they could actually go to a construction site and talk to somebody who didn't know how to read, and they were implementing some of those things. The barrier to change some of the mindset there was much lower at the construction site level because some of these people, they didn't know how to read and write, but once they were brought into the discussion, and were given some ideas, they would use it. If it benefited them, they would do it. It was a very big shock when I came from there to here and I had to adopt my teaching and the sites were not there for them to see, and people here, they seemed to be more reluctant to accepting some of these things. One of the most interesting things was that the terms that we were using to teach lean to engineers, some of them were not translated to Portuguese. When you talk about certain terms, the students had to learn what those terms meant in English. The term was in English or in Japanese or whatever language and they had to learn the term. I didn't think that that was actually a problem for them. You would just say what the look-ahead schedule is, and they would get that the look-ahead schedule was whatever I was explaining. They would call it as such in English.

FIRST THRESHOLD: BUILDING A LANGUAGE BRIDGE

When I came back to U.S., I started seeing that some of those terms that I was trying to teach back in Brazil and those students would just learn a term like "look-ahead" and they would make sense of it, when I got back to U.S., I realized that some of those terms were not commonplace here either. It was interesting to figure out how I was going to teach that because there was not this, let's say, inertia, right. The term "inertia," is translated into many different languages and there are formulas that are associated with it. Well, with what I was teaching, there was not. It was interesting when I came back here to see that those students were having trouble getting some of these concepts that I thought, "Okay, it's in their language, they are going to capture it better." That pushed me to be ever more creative in terms of how I was teaching these things. I had to become very creative as my other instructors were in the past. That's how I got into this track, if you will.

BELLY OF THE WHALE: THE TASK OF TEACHING

[I realized that] I have to adapt whatever I'm doing and see what kind of population I have here. SDSU is a university that is supposed to form people to go to the market, so they are not going to become researchers. Going to the Center for Teaching and Learning [CTL] lunches here, they said, "Remember that your goal is for students to learn. Do whatever you have to do, but they have to learn." I think that was that "aha" moment that I can do whatever I want, but these people are my clients and if they are not happy or if this is not useful for them, I have to do something. I have to find the happy medium, which I think I ended up finding after going to the CTL meetings and seeing the different approaches and using Blackboard and working on my syllabi to make sure it's clear and they know when the assignments are coming, when we have simulations and just keep reminding them. I had to become this person that serves more of the students and tailor my teaching to their needs.

ACCEPTANCE OF THE CALL: FOOD, HANDMADE LEGOS, AND PRESENTATIONS

I like food. I would always start talking about some food related stuff [in my classes]. I would catch their attention right away. I was talking about something that was very personal and I would say, "I like this, I like that, and now imagine that you are in a restaurant and that's what's happening." I would always try to anchor the concepts into something that they already knew and they were familiar with on a daily basis. To this day, when I teach, I talk about all the food places on campus. In one of my classes I used to send them to do studies in these places before I would send them to a construction site.

That was one way. The other way was the simulation with Legos. Probably you heard about many of them, and I created my own with, I asked the students to cut dice, small dice, and I would give very simple instructions and they would do it, and then we would move on to different problems of the game with them, making dice and making the airplane game that you might come across as you talk to some people in this field.

I remember a professor in Berkeley who used to bring some ingredients to class and mix in front of the students. He taught construction materials, and he would show how those things will add glue to the mix or become more watery or harder. Usually we see a lot of that in construction, but I don't see [it] in the other disciplines in my department, unfortunately. They might have other things that they do that I don't know, but as far as these Lego simulations go and stuff like this, it's just construction people.

In my grad class, which is the one that I teach these concepts the most, I created an assignment that they have to create a video or an animation or a game that explains a concept. They have to figure out, I just tell them what the parameters are. They have to pick a concept. They have to do a lesson plan. They have ten minutes to present and the video has to be up to five minutes within that presentation. It's something very focused, and they presented that.

ULTIMATE BOONE: POSITIVE FEEDBACK

A few weeks ago, and every time they have a presentation, I ask them to post on Blackboard a positive thing, a negative thing that can be improved, and a lesson learned. By far and large, they love this assignment because they could see the concepts, the abstraction that the metaphor for different concepts in each group presented in a different way. Some people presented washing dishes. Some people presented how they prepare to go surf. Some people presented how they are getting ready to organize a new production line in their company. The comments on Blackboard overwhelmingly tell me that they enjoy that because they could see different ways of applying the same concepts.

APOTHEOSIS: MORE PLUSES THAN DELTAS

[These comments are] usually very positive because we still have the traditional lecture [when] I'm in front of them and I'm lecturing the traditional way, but we have a lot of guest speakers and we have these simulations. Every time we have a guest speaker or we have a simulation, or they present, they have to do this plus/delta of lessons learned. This plus/delta lesson learned is open for everybody to see, so they see what other people write, and it's a safe environment, it's free of criticism. They post whatever they want and I put my comments as well, but they can all see what they're saying.

The only negative comment that I got that I would say for my video animation assignment is that they want more time to present. I'm saying, "No. I'm not going to let you get more time because you're going to be rambling there and people are going to get bored." That's the only negative thing that I have gotten so far. They want more time.

MASTER OF TWO WORLDS/FREEDOM TO LIVE: PEDAGOGICAL FLEXIBILITY

I feel that every time I'm going to lecture, I have my slides that are ready, but I never have the same exact lesson. Never, ever. Every time I go to a class, not only I have to see my notes, but whatever I'm listening on NPR, if I have a chance, or I read the news, I always try to bring something that is contemporary, if nothing else, just to catch their attention. They might be distracted and say, "Oh, did you guys see this and that?" This relates to the class. I want to keep using that and I want to flip the class a little bit more moving forward. It's very hard to come to grips with the idea that you want to introduce these things and you want to give freedom for them to lead the class, and at the same time cover the course material.

Right now, I hope I can better manage this struggle that I can just smoothly put these innovations in the class and still be able to move on with the content. This was another thing that the CTL changed in my mind before I wanted to go and bang, bang, bang, bang, cover the syllabus. Right now, I said, "You know what? If they learn this and this and this and they really know well about this, I can skip a topic or two and maybe have a smaller amount of time dedicated to that." I think that's my battle is to change a little bit at a time but still cover the material.

MEETING WITH THE ALL KNOWER: LIKE-MINDED EDUCATORS

[I also received support from the ASCE workshop.] That was the best work week of my life, because we would work from 8 o'clock [in the morning] to 8 o'clock, maybe 9, 10, 11 [in the evening], and we were all so happy, so excited, and that was another thing that I appreciate having a chance to go there and have this support that you are mentioning, because that's another thing they said. You have to think about ways to teach your students, and engineering professors are very, "You teach like this or teach like this." We went to West Point [for the workshop], and you had all these military professors that you would think are very structured. It's very structured, but it's also very fun. They had all these different things that they would do.

That was a huge help for me to be part of that and to accept that those things are accepted in engineering, because some people, I have the impression that when you say that you are doing these things, people think, "Are these people really learning? Is she really teaching something?" With my comments, my reviews, right, you will see that the students enjoy and that they like this approach. That was a huge help.

MASTER OF TWO WORLDS: LEGOS AREN'T A WASTE OF TIME

As far as the CTL help here, the director of the CTL here just invited me recently to be part of their first advisory board that they are putting together, so I'm having a chance to actually look at some of the schedules and workshops and things that are more hands-on. They are doing that and whoever is part of the board is trying to say, "Yeah, we prefer to have hands-on things that we can bring to class immediately."

I just wish more people [who teach engineering] spent the time to do these things, because I have the impression that sometimes when some of the students from other disciplines, when they land into my class and they say, "We are going to work with Legos," they are like, "Oh, there it comes." They think that it's something that I'm just doing to kill time, that I'm not prepared for class and then I'm going to play a game or something, and then some of this perception changes, and when I say we are going to play a simulation, and they know that it's serious and they will be engaged and I'm not just trying to kill time. I think that it will be good if more people had this mindset, that they could try and use these other experiments in class.

CHAPTER 7

Finding Her Niche with Hands-On, Practical, and Real-World Pedagogy

Fernanda Leite

Narrative constructed by Nadia Kellam

I think I'm still a work in progress. You said [that I'm a] rock star, but I just see myself as somebody that's just constantly learning, and constantly trying to provide our students the best education that they deserve to have. That just is a work in progress. Honestly, there are still days that I come out of a class and I said, "I could have done that better. Next time I'll do it better." It's always a work in progress. That keeps me motivated.

Fernanda Leite is an Associate Professor in Civil, Architectural and Environmental Engineering in the Cockrell School of Engineering at The University of Texas at Austin.

CALL TO ADVENTURE: COMBINING PASSIONS

My father is an educator. He's a professor in Brazil in agricultural engineering. I grew up actually in College Station, Texas. I knew what it was like to live the academic life from my observations of my father. I also was very passionate about construction, which was my grandfather's profession. He was a developer of high residential/commercial construction in Brazil.

My first teaching experience was teaching English as a foreign language in an afterschool program. That's where I fell in love with teaching, when I was an undergrad [in Brazil, where I'm from.] I knew that I wanted to teach, but I knew I didn't want to be an English teacher in an afterschool program as a full-time career.

I put all these little pieces together. Really, the passion that sparked that, that was ignited when I was just teaching. It was better than flipping burgers. I had the skills. Why not use that? The research observations of my father, and the construction domain from my grandfather. I wanted to combine the two [professions.]

SUPERNATURAL AID: FIGURING OUT HOW TO BECOME A PROFESSOR

[When teaching English in the afterschool program,] I just had lots of fun… just seeing people, observing people grow, and how a small intervention could really impact people. That, for me, was really encouraging, and it just gave me a high. The same thing that I feel after teaching a good lecture. Endorphins go off in your brain, or something like that. It just feels really good. [However,] I knew that that wasn't the domain where I wanted to be doing that, in terms of teaching English. I wanted to teach my chosen profession.

My father really helped me shape how [to reach my goal.] That's where I traced out my plan of, "What do I need to do to be this person, in terms of getting the right degrees?" I sat with my dad and said, "What do I have to do?" He said, "Well, if you want to be a university professor, you've got to have a Ph.D." That's where it started. From there, I went into a Master's program in the south of Brazil. I'm from the north-east [of Brazil.]

In Brazil, at least, at that time, I didn't know I was going to get a Ph.D. in the U.S., or become a professor in the U.S. My plan was more, "How do I become a professor in Brazil?" Because that wasn't in my radar, that this would be a possibility. My dad's like, "Well, you need a master's first, and then a Ph.D. Then you can apply for a faculty position in Brazil."

MEETING WITH THE ALL KNOWER: A VISITING PROFESSOR AND FUTURE ADVISOR

During my Master's [degree], a professor from Carnegie Mellon University went and taught a one-week short course over the U.S. summer, our winter in Brazil. At the end of that course, he basically said, "Would you like to get a Ph.D. at Carnegie Mellon?" I said, "Sure. Are there two positions, one for me and one for my husband?" Because my husband was also in the same path. We were both Master's students together. We ended up going to Carnegie Mellon for our PhDs, as well. He said, "Sure, I'll get a position for the two of you." We applied, and it worked out. We actually only applied to two U.S. universities for our Ph.D.s: UT Austin and Carnegie Mellon. We were accepted to both, but we decided to go to Carnegie Mellon in the end. [Eventually,] I ended up here [at UT Austin] anyway. Because I grew up here in Texas, so I had a big connection with the state.

ROAD OF TRIALS: EXPERIENCE TEACHING IN GRADUATE SCHOOL

I've always served as a teaching assistant in classes in grad school. I've always done research, which was my primary responsibility in grad school. I've always been really, really passionate about teaching. What I noticed [when I taught in graduate school] is that a one hour lab was pretty limited and it was, most of the times, very disconnected to what was happening in the three hours of lecture in the week. My Ph.D. advisor taught those, and then I taught the one-hour lab. There was that disconnect. That was the first thing. My desire was that it would be all connected, better connected.

I was frustrated [then,] because I didn't [teach in a hands-on] way as a graduate student, as a TA. I was supposed to teach the lab, and teach them how to press the buttons. That just frustrated me. But you only had one hour a week, and it was not connected to the lecture slides. You really couldn't do a lot more than that anyway. That's the first thing that I said. "If I'm going to do this, I'm going to do it right, the way that I really believe how this should be done." For me, since I'm in such a practical field, which is construction, I don't understand how I can teach without it being hands on, and very practical, and very real-world oriented. I just don't know how to do it a different way, honestly.

When I was a teaching assistant, I think the hands-on component was very limited. I tried to develop my advising style, my teaching style, based on my own experiences working with other people.

APOTHEOSIS: DEVELOPING AN INTERCONNECTED COURSE

When I came here to UT, that was one of the things that I created was this BIM (Building Information Modeling) course. The way I thought of creating those connections better was by dividing the course into modules. The modules would be a lecture, two lab classes, and then a reflection class. All on that same theme. They're all very interconnected.

The first lab class, we teach them, they're able to use five or six different software systems to be able to do applied BIM for different application areas in construction. There's a lot of new software that they're learning throughout the semester. Each module has at least one or two [types of software]. That first lab class is getting them up to speed in those software systems. Then they typically have one week between that lab and the second lab.

The second lab, I call it Time for Questions. There's no teaching component. We're not showing them anything. We're just walking around the classroom answering questions. Most of the teams have done about 80% of the work between that first lab and then the second lab. That way we're just really helping them connect what they've been doing to the general theme of that module, and answering questions.

I tell them that, "I don't answer button-related questions. Don't ask me a button-related question. Personally, I don't really care about teaching how to press the right button in the right order in the software system. For me, I care about what are you getting out of that decision support, that software system, the output that you're getting? How is that changing your decision-making process to solve that problem?"

I tend to focus more on the process of the module, because I really believe that they can pick up the software, wherever they go. Whatever software I'm teaching them here might not be the one that they're going to be using when they go out in industry. I really don't put too much emphasis on that. That, I think, is the main difference between how I teach them in class.

Over the summer of 2015, I participated in an academic BIM symposium, where faculty from all over the U.S. that teach BIM shared how they teach it. Most people tend to focus on ... One of the major softwares that [we] use is called Autodesk Revit. They said, "Oh, I teach Revit. How to draw a column, how to draw a wall." Well that's not really ... You're teaching them how to use a program. For me, that is a disservice to our students. Because there are tons of YouTube tutorials online that they can learn that through. You really have to teach them how to make decisions with those systems.

Then the second component is [that] it's got to be all based on real-world problems. See those large drawing sets over there? They actually use that. That's a commercial building in a different state, in Pennsylvania. That's what they use for two of their homework assignments in my BIM class. They literally walk around with those giant sets of drawings. It's a real building that has been built. Another homework assignment is another building under construction. The fourth homework assignment is this building we are in [at UT Austin], ECJ (Ernest Cockrell Jr. Hall).

They do different things with these real-world projects. That's important, because [the students] need to understand project complexity. In engineering, we tend to over-simplify problems and provide and spoon-feed a lot of the boundaries of problems in a way that, in the end, there's only one right answer. You give them all the assumptions that they're supposed to make, you spoon feed all of the inputs. That's not how it is in the real world.

When I show up in class with the first module and show them these drawings, and tell them homework 1, which is a model-based cost estimating assignment, that they're not going to find the specs exactly like it's stated in that project, in those drawings, in the specifications for that project, in the National Standard for Cost Estimating. They have to make assumptions, they have to interpolate, they have to find approximations.

Some students go crazy, because they are just not used to that world. [They say], "What do you mean I have to make an assumption? What is the right answer? What is the number that I'm supposed to get?" This is a cost estimate, it's an estimate. There is no one right answer. Everybody's going to have a different answer in the end, based on their assumptions. If two teams come back to me with the same answer, that's when I know there's a problem.

At the beginning, they're a little shocked in the first assignment of the semester, but then they get used to it. When they understand that, they really flourish in the class. I notice that a lot, especially with students that have had no internship, no industry experience, a lot of our undergrads are like that. This is a cross-listed course, with graduate and undergraduate students. It's about half undergrad/half grad. I see that reaction a lot with undergraduate students.

Each module is basically the same structure. Lecture, which is the theoretical basis for that module. We typically have a reading associated with that. The two labs, the first lab to get them jump-started; second lab, time for questions. Then the reflection class. Which is, if there are eight teams in that semester and four homework assignments, then two teams would present for every homework assignment. Each team has specific points to cover in their presentation, so that the presentations are not really repeated. It's more meant as a discussion. Everybody is expected to participate in the discussion and chime in, because everyone has had that experience. It's not like a project that only that [one] team did that one thing. They're presenting everybody else's, that's all new information, no. In this case, everyone had that same experience, so they're all expected to provide their input, as well.

There tends to be a lot of interaction and discussion in this class. Even in the first lecture-type class, I start the lecture, I just leave the PowerPoint up in the background, and we're discussing the assigned reading for that class. There's no PowerPoint, it's just the first slide is up, but I'm sitting there trying to get their perspective on what were the main take-home messages from that reading. I really try to cement the important concepts in that reading. Then we get into the lecture. Then I basically tell them what the structure of that assignment, that module, is going to be after that.

The reflection just caps it all off. We're able to provide some closure for that module, and then discuss: "What were the limitations?" "How would you do this differently if you had this other piece of information?" And so forth. There's a lot of discussion that happens. It's very, very interactive.

[There are] four modules that are structured that way throughout the semester, and then there are several guest lecturers. I'm very much a believer of real-world knowledge. I invite people to come and guest lecture to talk about how they're using BIM in the field. We actually have two site visits. We're actually going to observe people using BIM in the field.

We've already done this two weeks ago on our first site visit. This is a high-rise project on West Campus, a residential tower. We went and visited their BIM office to see how they're using the models in their field office, and then visited the job site to see here's what they did, talk about in the virtual world, in the 3D model, here's what they're doing in the field. When they actually see it, and see other people using it in the field, it really sparks their interest. All of these different perspectives help them cement these concepts. It's not like they're just getting a one-sided perspective, it's [that] other people are showing them how they do this as well. It's not just one true reality. People apply this in different ways. It's important for them to see that experience. Also, each team of students also has an assigned industry mentor. They work with that mentor throughout the semester developing a case study on a real-world project that uses BIM.

As you can see, all of these assignments, there's nothing that's very spoon fed, like assumptions. Everything [assigned requires that] you've got to go out and try it. There's some structure. There's some material. But you've got to make decisions on your own. You've got to understand that, in the end, you're the professional. You have to take ownership for your work. We're not going to spoon feed you. You've got to go after all this data.

RETURN THRESHOLD: BRINGING THE REAL WORLD INTO THE CLASSROOM

Before the semester started in January 2015, in December 2014, I held a brainstorming session with an industry group called Safety Community Practice. These are about 32 professional safety engineers in this group from all over the world. We had a brainstorming session, over Go-ToMeeting, on what the next generation of safety engineers should know. That's how I created the topics, the structure of the lectures.

It's not my domain, expertise, but I really wanted to teach that class. But I also wanted to reach out to people that it is their domain of expertise. That's how the lectures came about. I put together the syllabus. Then for each lecture, I thought about, "How can they do that reflection in the class?" Because I don't want to just read lecture slides, I want them to be able to reflect on their own, on that theme. For each lecture, I did something hands-on. It was either a case analysis, which happened a lot, and they really loved it. A real-world case and they have to deal with, let's say, an accident investigation. What were the different steps? How they would do an accident investigation for that case?

One day, we used the intersection of Dean Keeton and San Jacinto, right here, [outside] of our building. We considered that an active job site. Each time somebody crossed, J-walked, basically, that was considered a near miss. They basically classified how many near misses were in that intersection. We were doing behavioral-based safety, that was a theme. They were using a lot of concepts that we had learned throughout the semester applied to something that's not a construction job site, but you could, literally, think about it as a job site. Because you still have behavioral issues. You could think about a car as being a construction equipment, and pedestrians as being the laborers on the job site.

We also went to this job site here, the new engineering building, and we did job hazard analysis, a theme in one of the classes. We went and looked at a set of workers that were doing a specific task around a column, we just identified all of the hazards in the field, in person. I got several exam questions just from [a] picture I took of this construction site from my office window that they had to do an analysis on, of a real-world problem. Nothing is memorizing and applying, it's really understanding the problem, and how do you connect the concepts that we covered in class to that real-world problem? How do you do something that people in the real world, a safety engineer, is actually doing? Like an accident investigation, a hazard analysis, and so forth.

MASTER OF BOTH WORLDS AND FREEDOM TO LIVE: ENCOURAGING OTHER ACADEMICS

I think the biggest barrier when I presented my approach in the summer of 2015 to other academics teaching BIM all over the U.S., people are just shocked at the amount of work. I think that's the first thing. It's too much work. Because you have all these mentors, you have all these

case studies, you have these site visits, you have all these modules. It's five or six different software systems. If people just look at it from a distance, they are overwhelmed, because they're going to think it's too much work. I think people are scared of things like that. Most faculty want to be able to not depend on other people. "I want to be able to go to my lecture and just do my thing. If I have to depend on other people, then it's a bottleneck."

The type of support that I have found helps is a teaching assistant; I have always had one. The teaching assistant helps update the tutorial material for the lab, and helps teach the button pressing in the lab. Because if I had to update all of these, because the software systems are updated every year, every single year... I've got to make sure that the right version is installed in the lab, I've got to make sure that the tutorial material's up to date. If I were to do that myself every year, that would be a huge barrier for me to keep doing it this way, because it's just a huge amount of time. Frankly, I'm not even interested in that part of the process. I do all the other [parts] of identifying the real-world problem that they're going to be using for the assignment. Identifying the mentors, connecting them to the mentors. Frankly, updating the material, I think, is the barrier. If people don't have that same kind of support, it becomes overwhelming to teach a class like this.

I try to do something, a smaller version of this, in my required undergraduate course, that we have five different structures on teaching. I try to make it very problem-based as well in class, but I don't use any software systems, which also minimizes that barrier as well. Same thing, I have a real-world project—they do an estimating project. I pick some area around campus, so they can actually go and see it. Normally, it's little plazas like the Barbara Jordan statue plaza. They do a quantity take off and a cost estimate for that. Then they have a panel of judges that are all UT project managers that actually worked in the construction of that plaza, and they evaluate the student's work, the student's cost estimate. That mini-project is probably the highlight of that course. I still do it. That takes about two to three weeks in a semester. Very hands-on, and it's completely real world. It's much better than teaching them how to cost estimate using a very standard problem from the book. It's boring, and it doesn't really show them the multiple dimensions that go into the problem. It's too simplified. It's nice to get the students out of their comfort zone, to really make them think, and not just blindly apply things.

I'm the only one that [teaches in a hands-on way.] Because most people will say that, "Oh, it's a lot of work, because we have to get all those projects, you have to get those plans, then you have to get that panel of judges." Honestly, I don't think it's a lot of work. I think the benefit is much larger than the work involved. Once you have a structure in place, all I change every year is the project. But this general structure is pretty much the same. I know if you build relationships with the right people, you can rely on them every time you teach that class.

If you just plan ahead, people are happy to help. It doesn't become that overwhelming. The students really value that, and they really enjoyed something that they can say, "I've worked on that project right there." Barbara Jordan statue plaza, or the Cesar Chavez plaza, or whatever it is on campus. They're really going to take that and remember that for the rest of their careers.

Again, people will say, "Oh, that's too much work." It is much easier to just get a book, walk to a lecture hall, and just teach straight from a book. Yeah, that's easy. For me, that's not fulfilling, and that's not why I chose this profession. That wouldn't make me happy. I would feel very frustrated.

STORIES FROM MY CLASS: TEACHING WITH LEGOS

Last Thursday in class, we did a Lego exercise. I can send you the examples. Basically, it's like a 2D set of drawings, elevations, plans, section cuts, of a 3D model made in Lego. They are in teams of 4, and there are 4 colors in this exercise. Each student is a different color. They have to build that model in 3D, but all they have are the 2D drawings.

We timed them. We basically see how long it takes for them to build a 3D model. It takes them between 7–10 minutes, which is a pretty long time, if you think about it. But it's to make a point. There are about 30 pieces, 4 colors, 4 team members per group. They worked together, and they have to communicate their ideas to make sure they're putting their pieces in the right place. But also the fact that they have to interpret that 2D information, translate that into 3D, that's also the point of this assignment.

The second round of it is I give them a 3D perspective of another model. They have to repeat the same exercise. They still have the same colors, same number of pieces. Now they're able to build that in 55 seconds to a minute and a half. It really decreases the amount of time.

That's to make the point that if you're able to communicate in 3D, which is part of BIM, building information modelling, if you're able to communicate in 3D, your crews in the field that are building that will have a very clear understanding of your design, of what you're trying to build. They're able to work more efficiently, because they're not spending a lot of time trying to translate something in 2D, which was already translated from your 3D original idea.

We run those through, two simulations. Between each one, we reflect. We think, "Okay, what did you learn?" In the end, we reflect again: "What did you learn?" This whole process was just this exercise, and reflection on what they learned. Little things like that I sprinkle throughout the semester as well.

I do a Lego exercise in my required project management and economics class too. It's literally two teams of students, one on one side of the table, another one on the other side of the table. There's 10 Lego pieces, that start with the first person on each end. Each team has a die, they roll it. If you get one, that means that your productivity rate for that day was one unit. You pass one Lego piece to the person next to you. The person next to you rolls their die, and they get three. That means their maximum productivity rate would have been three that day. But since the previous person was a bottleneck, they can only pass on one. Then the next person...

We keep doing that exercise. Say the first person on the other side got six at the first roll of the die, so he or she passes on six. The second person got four. He passes on four, he keeps two in his station. All of the pieces that are left over after that round, then I ask the class, "What do they mean in the concepts that we covered?" Because the pieces that stayed in your station, they're called work in progress, in construction. The roll [of] the die of one, that's a low productivity rate, that's a bottleneck. If they're able to play that and see those concepts, it cements it in a much better way. It's a very simple exercise that takes less than 10 minutes with a discussion in class. It's more effective than me going through those definitions in a regular lecture. Then they're able to really see it, and experience it, and they tend to quickly learn it. And they'll probably never forget it.

MASTER OF BOTH WORLDS AND FREEDOM TO LIVE: INTEGRATING TEACHING AND RESEARCH THROUGH AN INDUSTRY FOCUS

I have my niche, which is, I'm very much connected to industry, even from my research as well. My department supports me. I can be productive and can build these connections. People respect me for what I am, because this is what I am, this is who I am. I'm just one person. I can't separate my teaching person from my research person. I'm one person. My experiences are all combined experience. That's what's important. I have to put in the same dedication that I do for teaching in research. That's the only way to keep teaching cutting-edge as well.

Especially teaching something that's very much information technology. That gets old really fast. You really, really have to be connected to research to keep students engaged, and what's the most innovative piece of it? Keep them ahead of the curve. That's the last thing. Just always maintaining that connection between research and teaching.

I actually have a student right now, one of Mary's (pseudonym) Ph.D. students. He's observing every single one of my classes for his research. One thing that he's probably noticing is that whenever I ask a question, out of the 21 students in the class, I see at least 7 or 8 hands go up. One third of the class, they immediately put their hands up when I ask a question, because they've had plenty of time to reflect on the question that I'm asking. I get to know all of them individually, so I tend to ask questions sometimes about their specific experience. They'll share that. They'll be able to communicate that well, because they've lived it as well. That's one thing that's amazing. You get a lot more participation that way, because they feel more confident.

All I know is that it's not going to stay the same. My hope is that a class like the one that I teach, the BIM class, is not going to be needed in the future, because it's just going to be industry practice. That's what I tell my students. My ultimate goal, my dream, is that I'm not going to be teaching this class in 10 years, because this is just industry practice, there's not going to be a need. I'm going to have to come up with something new that's going to be the next big thing in the industry. I'm going to adapt with time. Luckily, I have that luxury of being able to tweak things throughout the semester, between semesters, and think about new courses as well.

I think it's stimulating, also, to teach new courses, because it forces me to think differently, and to teach things in a different way. Because each domain has their own specificities that require you to adapt to that and try to deliver that material in a different way.

I think I'm still a work in progress. You said [that I'm a] rock star, but I just see myself as somebody that's just constantly learning, and constantly trying to provide our students the best education that they deserve to have. That just is a work in progress. Honestly, there are still days that I come out of a class and I said, "I could have done that better. Next time I'll do it better." It's always a work in progress. That keeps me motivated.

Doing that also makes me come up with ideas for research. Ideas that I have in research become modules, or lectures, in my course. For me, a lot of people say, "Teaching takes away from research. That's just taking up too much time. I don't want to do that, because I'm so busy with research." For me, it's all one thing. The way that I see it is that I build off of the experiences that I have in the classroom. That gives me lots of ideas for research, and vice versa.

CHAPTER 8

Creating a Community of Collaborators to Achieve Curriculum Change

Charles Pierce

Narrative constructed by Audrey Boklage, Brooke Coley, and Nadia Kellam

I want to share what I think engineering is, because that certainly has changed over time... Engineering is helping people. That's what I think of it as. We solve problems for [the] purpose [of] trying to help come up with solutions to problems that impact society. That's kind of by definition what we do.

Charles Pierce is an Associate Professor of Civil and Environmental Engineering at The University of South Carolina.

CALL TO ADVENTURE: TEACHING RUNS IN THE FAMILY

My dad was a civil engineer and also taught years ago back when you could teach with your Master's degree, which is what he had. He taught at URI (the University of Rhode Island) for a few years. I bring that up because I was aware of what his professional trajectory was, [and] most of his time was [in] professional practice, but I know he had some teaching experience. My mom was a nurse, which is important. I think in many ways, I'm a perfect blend of my parents, because I've got the technical side from my dad and my mom helped people. I had a pretty good idea that I wanted to go into engineering from high school, [and] into college. My parents never pushed, but I was well aware of what my dad did and I seemed to have those interests.

I earned a civil engineering degree from the University of New Hampshire (UNH). It was a reasonably small state program, which was a good choice for me, and I really got to know a lot of the faculty. I had some very important relationships with professors as an undergraduate student.

SUPERNATURAL AID: GRADUATE SCHOOL ADVICE

I was pretty involved at UNH. I joined ASCE (American Society of Civil Engineering) and became president and all those things. I do remember one professor in particular during my junior year encouraged me to look into doing a Research Experience for Undergraduates (REU) program. Of course, at that time, I knew almost nothing about [doing research]. I applied to an REU program at Cornell and was accepted. I did that the summer after my junior year.

As a junior, I already had a sense that I would go to graduate school. When the opportunity to gain research experience [was presented to me], I figured it was a smart move. Plus, at the time, [the market for] finding part-time jobs in engineering was really bad. Even with my dad being in the industry, he was unsuccessful at finding an opportunity for me to do an internship. It's not like I had other options.

I went and had the summer experience at Cornell University, which was very favorable. I got a sense for what graduate students did. I think that really set the stage for me to apply to grad school my senior year. I even went to one of my professors and asked him, "Hey, I think I'm interested in geotechnical [engineering]. Are there some programs that you would recommend for me?" Northwestern University was one of them and Purdue University was another where he also had a classmate. I applied to four schools. I applied to Virginia Tech, where my uncle was on the faculty in mechanical engineering, so there was a connection there. I applied to Purdue University, Northwestern University, and Cornell, of course, where I had gained the summer experience. I was accepted at all four and received funding from two.

I distinctly remember having a conversation with one of my professors at UNH. He suggested, "go where there's funding. You should be funded to go to grad school." That meant deciding between Northwestern and Cornell, which were the two that had made me offers. Ultimately, I thought Northwestern was a better fit for me than Cornell, partly because I had attended a fairly rural, small town school in New Hampshire. The thought of being able to go to Chicago and go to school there was appealing.

I visited all four schools my senior year to make a decision. I remember visiting Northwestern [and] meeting the faculty in the geotechnical program, which was my specific interest within civil and, more importantly, meeting the graduate students. Of course, I didn't know what to expect, so I'm like, "I'm going to meet all of these really intellectual, you know, above me kinds of students." I was pleasantly surprised to find that there were students I thought were very much like me, which was important in making that decision. That's not to say that [it] wasn't the same at Cornell. But, for whatever reason, at Northwestern that really struck me.

SUPERNATURAL AID: PUSH TO PH.D.

[I decided to attend Northwestern] for the purpose of getting a Master's degree. [One day after I had applied to] the Master's program, I received a phone call from a professor there. He said, "Hey, I saw your application. I saw that you applied for a Master's degree. I really encourage you to apply for the Ph.D., because then you're eligible for fellowships." I would not have otherwise been eligible for such funding. I remember thinking, "Sure, I'll do that. Why not?" [I'm unsure whether I had an] inkling to get a Ph.D. [at that time]. I remember really having to think about that decision, though I was intrigued by the idea of getting a Ph.D. My goal at the time was [to] get a Master's degree, be like my dad, go into professional practice. That was my intent. I now had a better sense of what a Ph.D. was, which I did not as an undergraduate, and could see what I could do with it, which was to go into academia. I enjoyed doing research, but never thought that was my strength.

I really did like teaching. I should probably rephrase that... I liked having good teachers. I liked being a student in classes where I resonated with professors that I thought helped me learn. I do believe much of that wasn't necessarily from my graduate program, but much of that was from my undergraduate program, because I feel to this day that I had some absolutely fantastic engineering professors in my undergraduate program. I knew very clearly, I could not go do that unless I had a Ph.D. I think [that] had there been more options, like it was back in the day with my dad, I might have stopped at the master's and then tried to pursue getting a teaching position. But I knew those opportunities didn't really exist anymore, in large part.

That was a big part of my decision making to get the Ph.D. knowing that I had to go through the research process. My end goal was I wanted to be a teacher. I think during my Ph.D. program, one of the things that resonated with me was having the opportunity to be a Teaching Assistant (TA). I think most of us that were Ph.D. students had at least one opportunity to TA. I don't think it was a requirement, but [being a TA] was certainly an opportunity.

MEETING WITH THE ALL KNOWER: AN OPPORTUNITY TO TEACH AUTONOMOUSLY

I was asked to TA the soils lab which is pretty common in geotechnical. Once I knew I was going to TA, [I went to the] professor who normally teaches that class and asked him, "What do I need to cover? Just tell me and I'll do it." [His response was], "This is your class. You do what you want." I was flabbergasted by [being given] that amount of responsibility. I was like, "Okay. All right, I can do that."

He just told me to run with it. I mean, he didn't hold my hand in any way, shape, or form. I don't think he even came down to the lab with me. He said, "You know where the lab is. Go find the equipment. Figure out what tests you need to run." I mean, he completely left it up to me. Whether he knew it or not, I have no idea, but that was the perfect thing to do for me, because it really did make me think about how the decisions about what to do in a lab class or any class [are made].

Just all the planning that's required for managing students and managing equipment and thinking about what you want them to get out of it. I'm sure I didn't do a particularly good job with that at the time, but I remember having to think about it. That was really important, because it confirmed for me that I liked doing those things. Given everything else I was supposed to be doing with research, I was spending too much time TA-ing. But it was important to me, so I did.

[After finishing my Ph.D.], I was interested in finding a place that had more of a balance with research and teaching. I knew I didn't want to go to a top tier research institution. I just didn't feel like that suited me the best. To be honest, I don't even know how I made that distinction. How [was I even defining] a good teaching institution? I don't know. I think it was more a process of elimination than anything else. Like, "Okay, I know that's not what I would consider a top tier research institution, so if it's not, then maybe it's more teaching oriented. I'll apply there." I was minimally informed back then in that process, but probably not as [informed] as I wish I had been.

I ended up getting an offer here at The University of South Carolina (USC), which seemed like a good fit. I had a sense from most of the faculty here at the time that teaching was important, and it was valued within the department. That was significant for me.

SUPERNATURAL AID: FUNDING SUPPORT

I came into USC knowing I wanted to be a good teacher. I [was also aware] of the research expectations and [knew that I'd have to] balance that. I was reasonably fortunate to get some grants funded early on. In fact, I think the first NSF proposal I wrote was funded. That, in large part, had to do with my collaborator. I was set up with a senior person living in Georgia at the time who had been a faculty member elsewhere and was looking to get back into academia. We were connected and wrote a joint proposal together. He was phenomenally helpful in that process in terms of learning how to write a proposal, which I knew very little about. I had some experience as a graduate student, but not enough.

I was fortunate to have had that. I was also able to get some local funding through the Department of Transportation. I was getting grants and [using them to support] students. I felt like I was doing the things I was supposed to be doing at the time, which was good.

ROAD OF TRIALS: NEW(ISH) CONTENT

I think the [funding success] allowed me to feel less pressured about the [research process] enabling me to spend time on teaching. I came in and was asked to teach a little bit outside of my comfort zone. I was asked to come in and teach a civil materials class with the associated lab. I had a little experience working with cement-based materials and concrete, which was part of this course, but I was not that familiar with a lot of the other content [without referring] back to my undergraduate days.

I needed a lot of prep time to learn [the course content] on my own and [be able] to share that with the students. Then I also had the corresponding lab, which I think was very good because it really forced me to understand material behavior. Not only did I have to teach it, but I had to be able to demonstrate it in the lab. I spent a lot of time working on that and then that next semester, I was [put in a similar situation]. I was asked to teach our soils class, which is a junior level class, and the corresponding lab as well. In some ways, that was good because it was the same type of teaching, just different material.

ATONEMENT: STUDENT FEEDBACK

[Starting off] at 28 or 29 [years old], I really enjoyed meeting the students, talking to the students, and trying to get to know them. I wanted to find out as best I could whether or not they were learning anything from me. There was one student who was also very willing to get to know me. During the middle of the first semester when I was teaching that civil materials class and the lab, I individually asked him, "Hey, how do you think the lab is going? Is this helping you learn?" He was actually honest and said, "Yeah, but I think you could do this, you could do that." [I found his critical feedback to be] great and was actually happy to receive it. It made a big difference for me by [enabling me to] understand [his perspective] of what was working or was not working. From that point forward, I always felt comfortable trying to solicit that kind of information from students. That was always an important thing for me, to try to get a sense from the student of what they thought they were learning.

ROAD OF TRIALS: COMMUNICATION IN THE CLASSROOM

Here I am, first year, teaching two classes, teaching two labs. I felt okay about that process. I know I worked hard in developing materials and strategies, although I don't know if that's really what I was thinking of at the time. [My conscious focus was more], "How do I try to get this information across?" The teachers that I liked as an undergraduate student were ones that I thought made the class engaging and entertaining. One of my environmental engineering professors, I particularly loved. She was hard and I did not do well in her classes, but I really liked her and I liked her classes, regardless of how I did.

I was not ever really good at having a plan for content to cover [during the class period]. [The plan consisted of] compartmentalized units of notes that [enabled me to know] exactly what was to be taught, [and] when. I quickly realized that [such structure] didn't suit me, because if students had questions on a concept that took 15 minutes of class discussion, I was okay with that. Many of them were still struggling with basics. I really needed to spend more time making sure they really understand the stress-strain curve, where the stress came from, how it could be calculated, and the difference between load and stress. I also took for granted students could make the connections on their own.

ULTIMATE BOONE: CONCEPTS NOT SCHEDULES

I realized I had to step back and make sure they understood the basics. Eventually, [I accepted that] if all I did was get them to understand those basics, [that would be] really good. As long as I feel like what we've covered they've learned pretty well, I feel like I've done my job. But it's still a challenge.

SUPERNATURAL AID: CLASSMATE ASSISTANCE

A former classmate of mine was also teaching a civil materials class that was more his background. We were classmates at Northwestern and he knew a lot more about cement and concrete than I did. I asked him for help with notes. He mentioned that he used this little exercise when teaching cement hydration. Cement hydration basically is looking at a chemical reaction between cement particles when they become in contact with water. They go through a hydration process and it's exothermic; there's heat released. Sure, I made sure I understood the reactions so I could explain them in class. But I remember thinking, "How do I get the students to better understand what's happening?" He shared with me that he used atomic fireballs, little candies, to illustrate that process.

Basically, it's a little exercise where you just go through showing several of the chemical reactions while [the students are] sucking on an atomic fireball. You associate where the fireball gets spicy to the heat release. Then once that wears off, you don't really notice it anymore, and that is correlated to a decrease in the heat released. I just thought [the exercise] was the coolest thing. I used it in class and I remember people loved it. So I'm like, "Yeah, that's a really neat idea. I need to do more of this, whatever this is."

ULTIMATE BOONE: CANDY AND PERSONALITY

I ended up teaching that civil materials class for a number of years, and over that period of time, I ended up developing a whole series of mostly food-related activities to try to illustrate certain concepts. I have one where I've [heated and frozen samples] of Laffy Taffy. When you pull on the heated sample, it really stretches, and so I tried to use that to very grossly exaggerate ductile behavior. [In comparison, the frozen samples] would become brittle and just fracture demonstrating that behavior. In doing a whole series of little things like that I recognized that students always appreciated when you did things that were a little bit different. Whether or not it was actually helping them learn, I don't know that I knew that at the time, but it seemed like the right thing to do.

I ended up trying to create a number of these kinds of activities in all courses I was teaching. Never in my graduate courses interestingly enough, but in all of the undergraduate courses I was teaching. I think having been a slightly above average student, I feel like that has dictated a lot of how I teach. I teach in such a way that I'm trying to reach everyone knowing that I won't necessarily reach everyone. I'm trying to really make sure students understand the most basic principles first and then build on those.

Maybe that was just going to be my personality in the classroom anyway, but it felt like, "I want to make it interesting and entertaining." I tend to talk fast and loudly. I definitely tend to not stand still, and that's just how I am. Even back then the students commented, "Can't you just stand still? I'm having a hard time following you walking around the classroom." [I'm] pretty animated in the classroom. That seemed to go over well, so that was reassurance for me, that I was at least approaching this the right way. Although, I feel like that was just my nature anyway. It might have been hard to change it. I don't know, but I felt like I was getting the feedback that worked well.

FREEDOM TO LIVE: FLEXIBLE SYLLABI

I wasn't teaching things like statics or solid mechanics, which are largely solving problems, doing calculations. I was sort of in this middle, more conceptual. The textbook for my civil materials course was very conceptual. There were certainly problems and equations, but not all the way [throughout the text. In the text] you're learning about concrete, what it is, what it is made of, and how you construct it. What is steel, what it is made of, how you construct it, so on and so forth. I think it made sense to me [at the time] that I had to do these sort of demonstration activities for them to understand the concepts. It wasn't just a matter of using calculations to solve problems, that was part of it. But they had to understand the concepts, too. I think that was the breeding ground for me, through what I now know as active learning, to realize that conceptual understanding was really important. I'd always been asking students questions in class like, "Do something. Do you understand that? Are you sure? Let's discuss it again." If I was asked a question, I would try to explain it in a different way if I could. I think it was always more natural for me to do that.

[I no longer] have a schedule in [the course syllabus] that says, "Day one or week one, do this. Week two, do that." With the way I teach now, I can't do that because I let my class evolve. Early on, my struggle was internal. I wanted to be that good professor that taught them everything. I would say, "Okay, no more questions. We got to move on. I need to cover this stuff." I was very good about covering material because I was fast.

MASTER OF TWO WORLDS: PROBLEM SOLVING IN THE CLASSROOM

I think what's happened over time is [that] I find [questions I pose during class] become more of the emphasis than a side note. That, of course, in turn, has evolved into doing things like worksheets associated with an activity. Not only do we do the activity, but I may have them complete a worksheet. Maybe that's individual. Maybe it's group. Where I am—on the conceptual side in terms of how things have changed—I do a lot more of those sorts of things. I also do a lot more in-class problem solving and I never used to do that. I guess this was somewhat initiated by the idea of wanting to flip a classroom. I got very intrigued by this concept a few years ago.

I like problem-based learning. "Here's a problem. I haven't touched everything about this, but dissect it. Here's the problem statement. What do you need? What's given to you? Ask me questions." That's what I like to do. I walk around in the room and get questions. Once I get a question that's been asked a couple of times, I raise it with the class. I love that. It is so much more suited for me. Again, I guess I have just gotten to a point where it's also [the] content I'm comfortable with, which I think makes all the difference. I feel like I should be able to answer any question they ask me, but I also feel like if they ask me and I don't know, I'll go try to figure that one out. Again, if it's in-class problem solving, I usually have a pretty good sense of what kind of questions are going to come up beforehand. I guess that experience helps me.

I'll pick a specific problem for a reason, because I know it's going to teach them X and Y when they go through this problem. Instead of me just writing stuff on the board, they're going to learn it by going through this problem. [I'll ask], "Does everybody understand the X and Y? Because that's what's important from this problem." I always try to make sure to emphasize that when we're done. I do a lot of that.

ROAD OF TRIALS: IMPROVING THE CLASSROOM

There was a group of us in the department, sort of like-minded, who enjoyed teaching but really wanted a better sense of what students were learning, [and] how to do a better job. We sat down to put together an NSF proposal for what was the Course Curriculum and Laboratory Improvement (CCLI) grant program at the time. The purpose of the grant proposal was to develop a new course. We wanted to develop an introduction to civil engineering course that we did not have [yet]. It was driven by wanting to improve the curriculum. We knew we needed a first-year course for our students to have a better sense of what they were getting into in the major.

That was the purpose, but at the same time, we knew, "Okay, what can we do different? What can we do to make this class more unique?" We didn't want it to be a class with a whole series of PowerPoint presentations on the different disciplines of civil engineering and off you go. We didn't want to do that. We really wanted to make it a more engaging class. I can't remember now exactly how we arrived at this, but we did want to introduce technology. More specifically, we wanted to introduce sensors. I think part of the reason [we wanted to introduce sensors] was that most of us on the proposal did experimental research, understood the value of sensors, and wanted to get that concept across to the students, thinking that [introducing sensors] would help them learn some of the things that we do in civil engineering, some of the things we measure, why we measure them, and how we measure them.

We were captured with this idea of critical thinking. I'm not sure that any of us really understood what that meant in the context of engineering learning, but we knew that was an important thing. I do think a lot of it was driven by [reflecting on] most of our classes in the curriculum, how students could go through calculations, solve problems and almost never think about the output—never think about the number, the value, does it make sense within some bounds of reasonableness, never think about the units... Of course, we complain about that all the time. Students don't really have a good understanding of units and how it's associated with the answer. One of my colleagues, I heard him say this in a class one time a few years ago and I loved it—I steal it and use it now—he explains that in engineering problems, the solutions have a name. They have a first name and a last name. The first name is the value and the last name is the unit. I just thought that was really cool. I like that because I think it really emphasizes that both are equally important and one depends on the other. You have to understand what units you're dealing with to make some sense of the numbers.

I think that was a big driver for what we wanted to introduce [to] civil engineering students coming into the major—how important that was. Ultimately, that evolved [into recognizing that] in order to do that, you almost have to think beforehand what would make sense for an answer to a problem. I think what we realized is [that] we never give our students a chance to do that [before].

We never ask them in advance, "Okay, here's a problem. What do you think would be a reasonable answer to this? Don't calculate anything. Look at it. Try to dissect it. What do you think is going to be a reasonable answer? What's the order of magnitude? Are you going to be in the thousands? Are you going to be in the thousandths? Like, where are you?" It's amazing when you start really stepping back and asking students these questions, how far off some of them can be. You realize, "Oh my gosh, I really need to help these students understand. Get a sense of what they're doing."

ULTIMATE BOONE: WITH FUNDING COMES CHANGE

I think we were fortunate to get that grant funded to develop that course. I think we offered it for the first time in 2007, if I remember correctly. That changed everything for me. Getting that grant, having to develop this novel course in a very unique, problem-based learning teaching style [was a significant opportunity. Additionally], working on that proposal with my co-investigators gave me a new appreciation for collaboration and what that could really mean.

I don't think we even knew that at the time to be honest. I don't know that we even used that terminology in the proposal, but that's essentially what we were doing. [We were giving] students these realistic engineering problems and ask them to estimate a solution knowing that they knew nothing other than any prior knowledge they might have had about how [to approach the problem] and solve. Our purpose was [to] get them to think about the problem, what are the factors, and what's even important. One of the problems I developed for the class was to introduce students to what geotechnical engineering is within civil engineering. I wanted to pick something that would resonate with the students. Again, this was '07. I picked a problem that was set in the context of Hurricane Katrina and the levee failures, because that was very recent at the time. I knew that was something that students would understand and were aware of.

Basically, through a long process we decided we wanted to ask the students a problem related to the failed levee section in New Orleans. It needed to be rebuilt and we specified the length. We said, "It's a 100-foot long section of earthen levee that needs to be rebuilt. How much soil do you need?" I specified in tons because we had a long discussion about assigning units. We said, "Okay, just to keep it consistent, let's ask for tons of soil," which is something that would probably be used in the field, thinking in those terms. "Okay, I didn't give you much other information. What do you need to figure out? First and foremost, draw me a levee. What does it look like?"

We had a little bit of discussion about what they were, but not a lot. I wanted them to see what they visualized, [which the research team] thought this was important, this whole concept of visualization. What they saw in their mind when they thought of a levee was fascinating stuff. You'd get some really nice two-dimensional drawings with dimensions and units as well as some abstract looking things that I wasn't sure what I was looking at. You'd get some that were asymmetrical versus symmetrical. You'd even get students who would draw three-dimensional drawings. Then you'd start to realize, "Wow. Now, you can start asking questions about why did you think about that or why did you choose this?" As part of that, getting to think about, "Okay, so you have a shape. How do you figure out the weight?" Keep in mind this was for freshmen. Really, we were just trying to introduce them to what civil engineering was and what we felt like was the process of engineering problem solving.

The whole purpose of that course was to have opportunities for the students to explore and refine their answers. Clearly, one of the things we realized very quickly is students can't really visualize a ton of anything. You asked them this fairly large magnitude unit, it's hard for them to think about. The other thing that was very interesting and that was dealing with earth and a levee. One of the things I wanted them to learn from this whole exercise was, "Okay, in civil engineering, when you're a geotechnical engineer, you work with building and designing earth structures using different types of soil." There are different types of soil from an engineering perspective. There are gravels, there are sands, there are silts, there are clays, primarily what we work with. The [different types of soil] have different properties, which is ultimately what they would learn later on as a junior, but I always wanted to expose them to that.

What was interesting was how many students were like, "Oh yeah, soil that's like the stuff in your yard with the grass, the top." They thought that you're building this whole levee out of top soil, organics and grass. But, it didn't occur to me that that's what a lot of students would think. You don't want to work with organics at all, so it was perfect. It ultimately [presented the] opportunity for them to figure out, and me to reinforce, why you would never want to build using these kinds of soils. They actually ended up working with some soils. I basically brought the four types of soils: the gravels, sands, silts, and clays. They would work with them and learn what the density was. They also learned what affected the density, which was the other thing I wanted them to figure out.

APOTHEOSIS: ENCOURAGING CRITICAL THINKING

One of the most important things for them to walk away with for me conceptually is that when you work with soil, soil density changes. It's not a constant. That for me was a huge change from the activities and trying to get students engaged. I realized that capturing what they thought they knew, like actually getting that written on paper, was so important. Until you get them to write the stuff down, you have no idea. It enabled me to look at it and see what they thought, giving me the opportunity to correct, or better yet, giving them the opportunity to self correct, which [was the intent behind] how we set up a lot of those exercises. From that point forward, it made me realize that doing worksheets and those sorts of things in class, not for grading purposes, [but for understanding], was critical. It's participation. I make that clear. "Do your best. Provide me your best set of answers. That's what I'm interested in. This is not for a grade," which in that freshman course, it actually works reasonably well because they're first year students and they don't know what to expect. We tell them, "Hey, this is how this class goes." They're okay, they're on board with it.

Ultimately, we didn't grade them until the very end documenting this whole process of what they'd learned. We essentially did [what ended up being] a report. "Write a report. Give me a final answer. Explain to me why you think that's a good answer." What we also did in those reports, which I really like, is explain the process of how your answer changed during that time period. Of course, some of them were better at this than others, but we forced them to say, "Okay, the first day I guessed it was one ton and then I realized that's not possible because I was assuming this, that or the other." Whatever they thought they knew. "Now, I know a better answer is, let's say, 5,000 tons." We wanted to force them to step back and think about what they'd actually learned through that process and document it. I think that process has become infused in basically every class that I teach now, [and every class I teach now] is using that approach. Active learning, worksheets to document [the learning experience], and in-class problem solving.

ULTIMATE BOONE: TEACHING AWARD AND REFLECTING ON MY JOURNEY

[My teaching has been recognized by my academic peers]. I was nominated for the Mungo Undergraduate Teaching Award, USC's highest level award for teaching, and I had to prepare a statement. I spent a lot of time thinking about that statement [for the Mungo Undergraduate Teaching Award] and [decided that I] wanted to share [about] why I got into teaching.

Maybe this was subconscious until that time, but it made me realize that, I think in many ways, I'm a perfect blend of my parents because I got the technical side from my dad and my mom helped people. I had no teachers necessarily in the family, but I think those two characteristics brought together sort of described me, because I do look at teaching as more than that. I don't necessarily want to be a good teacher. I want to be thought of as a good mentor. As a person who's there to help students in this whole process of being an engineering student.

I've been working more in the past few years with K-12 teachers and students, and part of the reason for that is I want to share what I think engineering is, because that certainly has changed over time. You know what? Engineering is helping people. That's what I think of it as.

We solve problems for [the] purpose [of] trying to help come up with solutions to problems that impact society. That's kind of by definition what we do.

MASTER OF TWO WORLDS: COLLABORATION IS KEY

I'm tenured, [but I value collaboration]. I almost don't want to write any proposal that isn't a true collaborative effort with people that I know are just as bought into the research idea as I am. [My colleagues and I] always put out a proposal I'm very pleased with, whether it's funded or not. I feel like we do a nice job. Man, I love that. It's interesting, because for me, that is so intellectually stimulating, which is what I think most people want out of being in academia. You have that freedom, that opportunity to choose how you want to be intellectually engaged. This stuff is fascinating to me because I feel like I don't know enough about it, so I want to explore it more. Having people to share those thoughts and not having a single concern whatsoever that someone is going to say, "That's a stupid idea." Someone may say, "I don't think that's a good idea and here's why…" I'm fine with that.

To be able to do this type of work on pedagogical strategies and curriculum change, [you want to be a force]. I cannot do this solo. What's the impact going to be if I do one thing in my class? The collaboration with the seven or eight faculty I've worked with has been the best part. It makes a huge difference for me personally to know that there's a decent size group of faculty that I brought into writing these kinds of proposals, to doing this kind of work, to [recognizing its] meaning and potential impact. That goes a long way.

I feel like I am constantly learning about learning. I'm at the point where I'm gaining some knowledge about student learning, how students learn, what's effective for them, but not nearly enough. I definitely feel like the next step for me is getting a better handle on how to assess that—how to really determine that what I'm doing is effective in the learning process for the student.

We've been very successful in developing the classes in the way we intended to and have collected data on that. Now, I feel like we have so much data that while we've looked at some of it, I need a better sense of how do I extract from this solid evidence of what worked and what did not work? It is interesting. I question myself all the time now when I do something in class. I ask, "I wonder how effective that was…" I still have tests to see what concepts they've learned or what kinds of problems they can solve. That's all good, but I want to know more about the process the students go through. I think that's where I want to move forward—getting a better handle on how to do those things, which I think should make me, and others, a better instructor.

CHAPTER 9

Teaching with Advocacy: Buffing the Talent to Break the Mold of the Monolithic Engineer

Matthew Fuentes

Narrative constructed by Brooke Coley

[I want to] empower others to do those things. I don't need my face on the cover of X, Y, Z. I would like my students' faces to be there, wherever they are, and to be representative of those talents that are really sitting around, and not being polished, if you will. Yeah.

So, I kind of think of faculty as more of park rangers, and this information as just kind of like the parks. It's not that faculty are less important. It's just this idea that there's this huge landscape of information that students have to navigate. They can consume it anyway they want, but it's really damn nice sometimes to have a park ranger around to ask those questions, to make those connections, to see stuff that maybe you would not have really looked for or at before. I think that's really the role of the faculty member, is that guide, that park ranger if you will, to this information.

Matthew Fuentes is an Engineering Faculty member at Everett Community College.

THE CALL TO ADVENTURE

When I was an undergrad, I was hyper-focused. I wanted to be an aerospace engineer. Nothing was going to stop that. This was what I wanted to do. It's sort of like when I started exploring in grad school, which is completely reverse of what most people do, it was, oh maybe I don't want to do, not that I don't want to do this, but I am interested in a lot of things now. So, I started exploring more, taking more computer science stuff, taking some more advanced math, and some algorithms, and just kind of going all over the place.

I think I got the hook for teaching when I was a tutor back in undergrad. I was a math tutor, [for the] math department, and that's where I started to really focus more on student-centered learning than just the faculty- or teacher-centered learning paradigm. That more one-on-one, walking you through the process. I'm a pretty social person, so I think that [the] social aspect of it—the human aspect of it—was what really kind of struck me at the time.

I guess fast forward a little bit. When I started teaching, I started teaching actually in graduate school when I was working on my Ph.D. I'm a Ph.D. dropout by the way, so you don't have to call me doctor or anything. I guess I realized I didn't know what I wanted to do, which probably scared me a lot. Because at the time I guess I wouldn't admit to myself that that was true, and I wouldn't admit that I wanted to change. The reason I left was not because I didn't feel supported. Actually, that's kind of an odd thing. I definitely felt supported. The reason I left was because I guess I finally recognized that I wasn't there because I wanted to do this particular research. I was there for the glitz and the glamour, and I didn't know exactly what it was that I wanted to focus on at that moment. And I really liked teaching, so why was I hyper focused on this if I wanted to teach?

I made the choice to move out West with my wife and just quit everything. I thought, "hmmm, I'll take a wild risk." I had never really taken a risk like that, so I'll give this up. Of course, friends and family were like, "What the hell is wrong with you? You're giving up your RA [Research Assistant] fellowship to go live in an apartment in Seattle?" [I was] like, "Yeah, but my wife will work at Microsoft, so we're fine, and I get to plan the wedding. That sounds fun." It was nice. It was nice to change. I think it was really good for me to make that change. I think it was hard at first for me to make that change, because I had always been the hyper focused, motivated ... I don't know, win with all costs comes to mind. It's like, you know, publish as many papers kind of person, to what am I doing again?

[I] started to reflect and [tried] to recognize what motivated me and made me really happy and appreciate things. I think it made me a better person doing that. One, it's fun to say that I'm a dropout from college. It at least starts an interesting conversation, but I think it helps me sort of make peace with the fact that I didn't need to be in that role to do what I loved to do. I didn't need to be a tier-one research faculty leading a research team, especially since I thought there was a lot of really talented people that didn't do that, or really didn't get that opportunity. I started to feel like, well I wonder how much of my success also is because people trust me because I look like a typical engineer?

SUPERNATURAL AID: LEARNING TO TEACH IN A STUDENT-CENTERED WAY

So, when I was working on my Ph.D. I really liked the teaching aspect, and my advisor at the time, he was pretty big into engineering and pedagogy. In particular, [he was big on] bringing things into the classroom to make it more, I guess, student-centered. More hands-on was his real approach. What was kind of interesting about that experience was I was his student at one point in time, and then I was kind of his colleague at some point in time where I was in the classroom with him. I kind of saw both sides to it, which was kind of a cool experience. The class that I got him for, and that he was really interested in, was Mechanics of Materials. [At] some places they call it Mechanics of Solids or Solid Mechanics. It's kind of, I'd say, one of the more visual of the engineering courses. It's pretty hands-on. It's pretty visual. [And], it's a pretty old part of engineering. His take was, "Well, why are we teaching such a hands-off methodology, you know, this lecture based [approach]? Let's talk about these problems, and let's really bring in some tangibility to this." That's probably where I started to really switch [my approach]. I not only liked teaching from the standpoint of bringing stuff into the students, but I liked learning in that environment as well.

I think the way that I have approached it is more along the lines of, in today's world, faculty aren't this big ball of knowledge anymore. You can find that information anywhere in the world. It's accessible by anyone. So, I kind of think of faculty as more of park rangers, and this information as just kind of like the parks. It's not that faculty are less important. It's just this idea that there's this huge landscape of information that students have to navigate. They can consume it anyway they want, but it's really damn nice sometimes to have a park ranger around to ask those questions, to make those connections, to see stuff that maybe you would not have really looked for or at before. I think that's really the role of the faculty member, is that guide, that park ranger if you will, to this information. I think making them more self-sufficient and self-reliant is important for when they get out into the working world and get to do their own things. They become lifelong learners.

THE CALL TO ADVENTURE: ASPIRING TO TEACH STUDENTS WHO ARE LESS PRIVILEGED

So, I started teaching Physics at a community college. Why did I start teaching at a community college? I think for me one of the other big light bulb moments was recognizing that not everybody's educational experience was smooth. Not everybody had the same opportunities as I did—a middle-class white man going into engineering—people kind of expected that of me in some ways. [As an example], I met a guy in the computer lab in the middle of the night, [who would later end] up becoming my best friend, and he was struggling with some stuff, with some programming. I ended up helping him out and chit-chatting. Long story short, what I sort of came to realize from him was [after spending] three years at a community college, he transferred to the University of Tennessee, where I was, to finish his aerospace degree. [He] now works for NASA. The part of the story that really stuck out to me was when he started, he was an auto mechanic and he told the guys in his shop, "I'm [going to] work for NASA as a rocket scientist." Of course, they thought he was a little crazy. What I really appreciate about his story was starting from essentially [the] pre-college math level, and then becoming essentially an active rocket scientist at NASA. That's what he does now, and we actually collaborate. And that's the kind of opportunity I wanted everyone to have.

And so, when I realized how much of a difference faculty really made in his life and him transitioning from that world into university, I sort of changed my focus on going to more of a four-year school to, all right, what can I do to be in this, I'll call it the transitional college—the open enrollment colleges? Universities know [their] baseline, student-wise, because you have entrance and admissions processes to go through. [But], what about all those other people that want to get to that point?

That's where I really—I kind of [decided], 'You know what, I should really try out this community college thing,' and so I started teaching at a small school. [I] started teaching physics at Cascadia Community College. I think they hired me mostly because they had an emergency fill. Let's see, I was hired a week before classes started, and it was courses that I was pretty familiar with, so I was ready to go. It was a pretty easy thing for me to start up. I think it was really in Cascadia that I guess I experimented a lot with different styles of teaching, and moving into the, how can I best empower my students to be these self-centered learners? How can I get my students to be empowered?

SUPERNATURAL AID: A MENTOR WHO HELPED ENCOURAGE EXPERIMENTING EDUCATIONALLY

What I really liked about the college that I started in was Cheryl Barnes (pseudonym)—the faculty member that recommended me—was actually the person that hired me. She's the type of person that's sort of like, "Yeah, try out whatever you want, go for whatever...experiment." She's very much into experimenting educationally. I think [her influence] and that experience just really helped me grow into an I-can-do-whatever-I-want [believer], teaching wise. Let me try some stuff out. Let's see how students react. What I really liked about what Cheryl did was she used the Physics by Inquiry, I think is the name of the little textbooks. It's really this motto of getting students to, in some ways, answer their own questions. You can ask probing questions, get them to work in groups, and get them to sort of discover all of these nuanced ideas to make the "aha" connections. I think growing up academically in that teaching system with her gave me the framework to start branching out from that. Because that [was] physics, but engineering has been very traditional. And so, it brought out more of the, "huh, well I wonder what kind of things actually do work in engineering?"

STORIES FROM MY CLASS: GOING OUTSIDE TO BRING OUT THE INQUISITIVE MIND

And so, I started doing strange things. When I say strange, [an example is], I took the class outside. I taught Mechanics of Materials in the spring and it wasn't raining here. So, I went outside, I brought some sidewalk chalk out, and we had lecture on the sidewalk. Part of the fun was we got [to be] outside. But, [also], there was some interesting spirit of literally walking through the steps, because you could write down a problem, have it flow, and really make students walk through the steps of a process.

They couldn't fall asleep standing up, so that's a good thing. It actually, what I found was, it somewhat brought out this, I don't want to say childlike experience, but kind of the inquisitive nature of the child mind, like this "oh, huh, I wonder what would happen with this," when we did that kind of experiment. It took them out of the classroom, "I'm just going to be absorbing knowledge brain," to "huh, okay yeah I can do this!" That was kind of one [part]. I think another part of that was bringing engineering from an enclosed room [to] out in the open, too. Maybe in some ways to socialize engineering and engineers.

My real hope was that passersby's, which sometimes happened, would just kind of stop, and listen, and be interested in engineering, and ask students questions, and the students would answer questions. That's my utopia that didn't quite take shape, but it did have some strange impact in that other people would notice, and they'd say, "Oh, that looks complicated." Then it would be an opportunity for me to say, "Well, you know, if you take this path and learn these things, it's interesting. It might be complex, but you can, you can totally learn this, too."

I don't have any data to say it totally reshaped all of these mindsets. But, I did like the camaraderie it created with my students. I did notice that they felt a whole lot more comfortable when they saw me doing these kinds of strange things to ask questions—to ask questions maybe they had been afraid to ask before.

ROAD OF TRIALS: FINDING A FACULTY POSITION AT A PLACE WHERE I CAN MAKE A DIFFERENCE

I was associate faculty at Cascadia for over three years, and they didn't really have the funding to put in a new faculty position. We were right in the financial crisis of 2000 something or other. I don't remember the date now. The state had frozen the budget, and so they had a hiring freeze for a few years. By the time they actually did have a position open, there was a position open at Everett, which was just north of us. I started teaching at Everett again, hmmm, this is a theme. I got a phone call over the weekend and a faculty member was pretty ill in Everett, and I had made some pretty good contact with the faculty up at Everett, and they were like, "Uh, so Matthew, we know that you teach these particular classes down at Cascadia. Is there any way you could come fill in midway through a quarter on a class? Because we trust that you could do that." I'm like, "Wow, I appreciate that you trust me." And, "Sure, why not?" I did, and it worked out. One of the reasons that I left Cascadia and went to Everett was really the students I felt like they were—the word raw comes to mind or scrappy.

So, Everett in many ways is a pretty rough city. Maybe in the news you've seen the city of Everett actually sued the drug company for the opioid epidemic, so we have a really nasty epidemic in Everett, and really—Snohomish County, which is the county that this particular school lives in. One of the kinds of interesting things is almost all [of] the students there start out in pre-college math, pre-college English, so not very prepared students. I guess in many ways they are the students that represented my best friend, the kid from Flint. It really kind of felt like that same group of students. The rough and tumble group as I call it sometimes.

I do feel like in many ways the program was a diamond in the rough, too—the engineering program out there. It was one faculty member when I started. No, I guess it was two, two faculty members when I started, to now we have five tenure-track faculty. Some of us are tenured, some of the others are still tenure track. And four associate faculty, so we're a really big department now, and just becoming this kind of center of hands-on engineering education.

ROAD OF TRIALS: BECOMING TRANSPARENT ABOUT WHO I AM

In one of our intro engineering classes—so one of the things that my colleague and I did when we first joined Everett was—we completely changed the first-year experience. We made it a lot more hands-on, a lot more tiered, so that we're taking you from wherever you're at and getting you up to sophomore-level engineering courses, in theory of course. But, not just we're going to spray you with a bunch of knowledge and expect you to grow from there.

I appreciated the challenge. It definitely pushed my teaching limits I would say, going to Everett. I don't want to say the students were less receptive of my quirkiness, but they were, I guess, more suspicious of it. Who is this weird guy? It felt like it took longer for students to buy into sort of my oddball schemes. Yeah, and I think I had to adapt a little bit, too. I had to better understand that a big chunk of my students in Everett are on the verge of homelessness every day. I guess the point is I had to academically grow, and change with that group, and meet whatever that need was, but still maintain this idea of empowering, and student-centered learning, because it felt like that was one of those things that could help a lot of these students out. I think a lot of students at Everett just, I don't know if they're not good at reading between the lines, or if they're not ... They don't know, so it's really good for you just to tell them what it is you're thinking, and why you're doing stuff. Because a lot of them don't have experience with college, or really very good experiences with the education system at all. I guess I became a lot more honest and open about who I was.

STORIES FROM MY CLASS: HELPING STUDENTS OVERCOME IMPOSTER SYNDROME AND BECOME MORE ENGAGED

I usually talk, so that brings me to the whole conversation of, "Hey what I expect in the classroom is to have conversations. If something comes up, you need to talk to me, we need to communicate. We're a team here. I know that all kinds of things happen throughout the quarter. The big thing that you can do is communicate with me. Let me know if something's going on, this, that, or the other because I'm here to help you." I give them the speech about, this class is all about you learning something new, so it requires that you ask questions, don't be afraid to ask questions.

I have a little stamp card that my wife and I came up with, a system. She's a professor as well. So, we came up with this little participation stamp card, and I give the spiel to the students like, "Oh, you need three stamps on this card." It's like a frequent coffee customer card is what it looks like that I put pictures of bunnies or my pets on, so they already know I'm a little different, right, when I'm giving them these. So, I give them the card and I say, you need three by the end of the quarter to earn your participation points, and anything above the three you can earn potentially bonus points on assignments. I explain to them I really want a way to incentivize you participating. It's good for me and for them because it's something tangible for the students that they can sort of validate themselves that they participated—that they are doing more stuff. It's also a way for me to point at something physical to say, "Hey, yeah, um, you want, you want me to go out of the way to help you with this, yet you really haven't shown a lot of effort in the course. You know, what gives here?" And now, I have sort of a physical record of what you've been doing or what you haven't been doing.

I think they appreciate maybe that I'm just honest with who I am. At least that's my hope, that they recognize that you don't have to be afraid to be an individual, and you don't have to be afraid to learn. There's no stereotypical type of person that needs to be in this classroom.

So, it can come around to the maybe they have the imposter syndrome, or something like that, then we can have a conversation. I think it's that aspect and those conversations—really directed conversations with my individual students—that have helped me transition, I would say. I think even if I were to go back to Cascadia and teach again, I would probably take all of that new stuff, and directly apply it to Cascadia as well. Yeah. Now I think I'm a little bit more focused on what I want as outcomes, and what I want for them in terms of helping them develop.

STORIES FROM MY CLASS: TEACHING THROUGH MAKING AND FAILURE

A good example of things that I think work pretty well is in the last of our series of intro classes, it's really kind of a maker's class, so all the students get a kit of sensors, an Arduino, and we have a 3D printer in the room. We set up the teams to be essentially, I would say, startups. They build two prototypes throughout the course, so two projects, and along with those two prototypes they learn how to program. They learn some basic team building skills. They learn how to problem solve. What I like about that class and what I think works pretty well, is giving students this freedom to search, try, find, explore, experiment, and have things fail. Failure is totally an option, and I think a great way for them to learn [is with awareness of] how hard sometimes things are to implement. Of course, you don't want it to be so much of a failure they learn nothing and they sort of unlearn everything. But, if I can, in those projects, get them to at least keep trying stuff—and to get stuff to work pretty well, and to have some failures along the way and fix those failures—that's the success.

ROAD OF TRIALS: INTRODUCING Simulink® BEFORE IT HAD BEEN DEBUGGED

One of the things that was a total disaster in that class early on was we had this—well there were a couple of disasters that happened. One, well, I think it was my doing here. I thought we were going to do this awesome thing. We were going to teach them MATLAB®, but I was going to teach them how to use Simulink®, and we were going to have the hardware software interaction. It was going to be amazing. [Instead] it was a disaster. Because Simulink®, at the time, had just unlocked the capabilities for Arduino, and I should've talked to the company beforehand, before making that decision to have students go down that route. There had to be a lot more debugging that was advanced for them that they couldn't really handle with the tools that we gave them. It wasn't such a seamless experience for what we were trying to have them do. I guess early on it was a reminder to me that something that was important for students in the early part of their career was having something that was challenging, but didn't sort of make it seem like it was impossible. In this, I think, early iteration, my guess is that some students sort of felt like some of the robotics stuff was impossible, right? Unattainable. That was not the message I wanted them to get. So, like, crap, I made this worse, great. I tried to contextualize it and say that, "This class is an alpha prototype," you know at the time, "and we're going to do some things that probably won't work, and let's just play. Let's see how it goes." I tried to remind them like, "Yeah this was terrible. I'm sorry." I think they got okay with that. They got through that.

ROAD OF TRIALS: UNCOVERING BIASES AND EXPECTATIONS AND A NEED FOR ENGINEERING TO CHANGE, CULTURALLY

I used to get interesting feedback all [of] the time in my course reviews, was something like, Matthew is really approachable, blah, blah, blah, awesome. His tests were crazy hard. He expects a lot. What I was getting in this feedback was students felt blindsided by, I'm a really relaxed person face-to-face, but not technical competency wise. That was hard for me to deal with, especially at first. I was like, wait, being nice and being technically competent are exclusive? I don't understand. It started to click a little bit more I think when I started seeing what my wife's experiences were. Interestingly enough, we've taught the same class before, in the same quarter. She would get dramatically different [evaluations]. I think one of the interesting things that happened was seeing how her students, I guess, expected her to be nice, like personality wise [because she was a woman].

Seeing how students reacted to her, because she's definitely firmer than I am, I guess what I noticed was, if I were firm or looking at some of my coworkers who are pretty, I don't want to say they're rough, but they're very strict—they have very strict schedules, they have a very strict classroom, style-wise—students never complain[ed] about it. But, when someone like my wife is strict, she's not really that strict, but she runs the classroom in a different way than maybe I do, the students react, I'll say, quite negatively to that personality type. I guess what it's kind of started clicking in was students' expectations on things like the gender role of the faculty member in charge of the classroom.

When I back up to that mismatch of me face-to-face versus the technical competency or capabilities on exams—back to those first comments—I guess I started to recognize that it was some kind of me peering into their biases or expectations. I don't have a good answer for that, but I think it helped. Those comments happened early in my tenure process at Everett, and I think that's where I started to become a lot more upfront with who I am at the beginning of the course, and lay it out, and not be afraid to talk about things if it arises. It's just that I feel like the world is changing, and engineering needs to change, culturally.

ROAD OF TRIALS: EXPERIENCING MARGINALIZATION THROUGH A LAST NAME CHANGE AND BECOMING AN ADVOCATE

I can't lie, having seen the world through my wife's eyes a lot more, she was at Microsoft for five years, and I saw indirectly how people reacted to her. I guess it sort of broke down my utopian vision of everybody lives with kitty cats and unicorns, and everything works great, right? [I went from thinking] there [are] no real problems, to wow, there's a lot more to this than I ever realized. There's a lot more to what people deal with on a day-to-day basis then I realized. I think the other part of that picture, of changing our expectations for what an engineer looks like is even a small part of what I experienced in changing my last name. Until I changed my last name, and I guess I started to feel—a couple of experiences. It's not uncommon that people ask where I'm from. I've had people say, "Wow, you're really white." Like, no one has ever commented on my race. What the hell? It was such, I guess, a shock. I really didn't expect that. For 30-plus years of my life, I had never had any comments like that. Now to suddenly, by a switch like that, [be exposed to people], I could see why it was annoying and why it was qualifying. It was sort of that kicking in like, well now, are you going to qualify that? I mean like now do I become a Hispanic engineer instead of an engineer?

I think it was really eye opening for me. It was kind of like trying on something new. It's like putting on a new face almost. I never anticipated having those experiences in that way, and being so directed. That's what was so shocking. But I think honestly looking at myself, I don't think I fully understood it, until I experienced it. While I can take that with the knowledge that I have of going through life I guess in a different skin, if you will. I can parse that now because I kind of know better. But, I can't imagine going through the educational system and having to deal with that and learning new things all at the same time. I guess to bring it all back to the classroom environment, it's that kind of conversation that I want to have happen in engineering classrooms, because I think those are important conversations along with the technical. Now, I know that you can't spend all the time in the classroom talking about these experiences, but it's important for students to at least be aware. For a long time, I struggled with, "well, all right yeah, but what do I do?" I really thank my wife for saying, "Yeah, but you have a unique role as the stereotypical white male in engineering to be able to stand up and do something about that, right? You can be an advocate."

I've tried to do a lot of soul searching on [diversity and people assuming everyone else is like them]. Quite frankly, I was thinking back to one of my roommates in college. He was also an engineer. He ended up quitting. [He is] African American, he was my neighbor at one point in time and we were really good friends. Now I go back to that a lot and think why was he not successful? What was it? I really wish I would have just asked him the provocative questions like, "Hey man, what's going on? You feel alright? Something going on?" Maybe I just am forever guilty. I guess I still always come back to I think people just don't have enough real conversations. They don't ask about how certain actions or things might make someone uncomfortable or something of the sort.

I don't have an answer for changing [that]. But, I at least want to have those conversations in the classroom. So, it may be over the course of my lifetime the needle is moved, right? If we make some progress, then we're stepping in the right direction. Maybe it's some part of me like, "Oh, he took his wife's last name, hmmm, maybe I should think about women differently, or relationships differently. He's interested in other cultures…" You know, maybe some part of my spectrum they'll at least take away from the classroom and maybe change a little bit, I'm hoping. Plus, being technically competent.

APOTHEOSIS: EMPOWERING STUDENTS

I recently met with a few of my former students from those—this is kind of an interesting aside—some of my former students that I had during that timeframe at Cascadia. And quite a few of them [now] have advanced degrees. One of them is finishing up his Ph.D. in civil engineering. Another one I just spoke to a couple weeks ago, he finished his Master's in four quarters in mechanical engineering. I think the great thing about these students, in particular, was that they were not very engaged. They didn't know what they wanted. They were pretty sloppy with their work. If you looked at them as a snapshot in a particular quarter you would say, "Yeah, they're not going to be successful." What I really liked was that they … I don't know how much a part I was in this transition, but at least I appreciate Cascadia being a part of this transition in them to let's say develop. Yeah, so I think that was just kind of looking back and seeing how I was as an educator, and seeing where those students ended up, yeah it feels good.

[I want to] empower others to do those things. I don't need my face on the cover of X, Y, Z. I would like my students' faces to be there, wherever they are, and to be representative of those talents that are really sitting around, and not being polished, if you will. Yeah.

CHAPTER 10

Conclusion and Lessons Learned

Nadia Kellam

I hope you have enjoyed reading these stories of engineering faculty and their diverse stories of embracing active learning strategies. To me, these stories highlight the complexities inherent in stories of change. Of these eight stories, there was not a single one that was simple and straightforward. This was part of the impetus for sharing these stories as people are not born good teachers; it requires work to become good teachers. While these stories show the difficulties in becoming exemplary engineering educators, they also highlight the benefits of changing our ways of teaching.

As I was reading through the chapters I noted some emergent take-aways that strongly resonated with me. This is not meant to be a comprehensive list of lessons learned. In fact, I am very interested in lessons that struck you as you read through the stories and anticipate that people at different points will appreciate varied aspects of these stories.

LESSON 1: IMPORTANCE OF HAVING A COMMUNITY

A lesson that appeared throughout many of the narratives is the importance of having a community throughout the process of improving your teaching. Sara reflected explicitly on this in her narrative when she acknowledged the support of her teaching-focused group of all women where she was able to hear the accounts of others suffering through the same challenges and feeling assured that, "Oh, it's not just me, I'm not alone." This larger network of women enabled her to not only learn content that would aid her in her pedagogical approach, but also serve as a critical body of support that would empower her to "get through that" period of her journey.

Charlie also described the importance of collaboration and community in his journey. Much of his personal satisfaction came from, for example, writing collaborative proposals with his colleagues in support of their teaching. He found that to be intellectually stimulating. He also discussed that making changes in a silo would not likely prove effective in achieving the scale of changes possible within an engineering program; Charlie wanted to have impact and learned early on that to do so required a community of like-minded individuals.

How can we all incorporate this lesson into our teaching journeys? One way is to find others who are also interested in improving their teaching. You may find some people like this within your department or program, or you may have to look more broadly to find others to work with so that you can inspire each other to continue to improve your teaching.

LESSON 2: THE POWER OF REFLECTION IN IMPROVING OUR COURSES

Sometimes as faculty we get so busy that we do not take a step back to reflect. As we all embark on our journeys in becoming better engineering educators, we can learn a lot through reflecting on our goals for a class, how the class went, how the students responded to the class, and considerations of ways to improve it in future semesters. Donna recommended taking time at the end of the semester to reflect on what worked well, what did not work so well, how you have progressed toward your goals, and, finally, what you will change moving forward. She describes this as keeping "the spirit of innovation alive." Most engineering programs have ABET requirements for continuous improvement, and this can be a good opportunity to reflect on a class and develop goals and ideas for future iterations of the same course.

How can we all incorporate reflection into our courses? There are many ways this can be achieved, both individually and collaboratively. One way is to write up your reflection and include it with your course files so that you can read through it when planning for the class in the future. Another idea is to have a discussion with peers in your community so that you can spend time sharing experiences from your class, what went well, and what did not go so well. This way your community can help you brainstorm ideas of improving the course or be able to anticipate areas they may wish to invest more energy into if they are to offer that same course in the future. While it is always helpful to reflect at the end of a course, it can also be helpful to reflect as the course is happening so that minor corrections and improvements can be made in real-time throughout the semester. Deploying a course evaluation while the course is happening can help provide some input from the students in reflecting on the course and to identify some opportunities for improvement. Another option is to ask a colleague to observe your class so that you can have another person's perspective on the strengths of your classes and areas for improvement.

LESSON 3: TAKE IT SLOW

In many of the stories, the engineering educators discussed taking things slowly and not changing everything at once. In Donna's story, she made some significant changes to her thermodynamics course over 10 years that made it "unrecognizable from the original one." Even though there were large changes over a longer time period, she explains that she "never completely overhauled the class." She would make one or two changes each semester. She explains that by changing one thing at a time, it did not become so overwhelming. By taking a slower pace, you are probably more likely to continue on the journey of changing your teaching. If you change everything at once, you could easily become overwhelmed and the teaching "experiment" could end.

How can we take it slower in introducing innovations? When planning our classes, we can think about our vision for the class and one or two things that we can do to help realize that vision. Through the stories, we learned that starting with one small change can initiate a series of beneficial changes. What are the one or two changes that you can make to your class next semester? How can you learn from those changes for future iterations of the course?

LESSON 4: IMPROVING TEACHING AND LEARNING IS A LOT OF WORK, BUT IT IS FULFILLING

Many of the stories highlighted in this book admit that their journeys of becoming better educators is a lot of work. In particular, Chris, Fernanda, and Brad describe their journey as difficult. While they felt that this journey was a lot of work, there was consensus around the idea that it was worth it. For example, Fernanda talks about how it would be a lot easier to teach directly from a textbook, but for her it would not be fulfilling. She explains that if she took this easy route, she would find herself frustrated and not happy in her position. She also talks about how she is constantly learning herself. She explains, "Honestly, there are still days that I come out of a class and I said, 'I could have done that better. Next time I'll do it better.' It's always a work in progress. That keeps me motivated."

Brad also discusses this idea that trying new things is a more difficult route, but that it is worth it. Brad explains, "Any time you try something new, outside the box, it's going to take a lot more time, more time than you probably anticipate." It will take a lot of time and energy, but "if you stick with it…it really does pay dividends."

What would our roles as engineering educators look like if we took the extra time and energy to become better educators? Would it make our faculty roles more intrinsically satisfying? These stories serve as an inspiration to try new things and continuously improve our teaching and student learning in our classes. They also help us see the importance of getting feedback, both formally and informally, from students to help improve the learning experience in the classroom and also serve as extra motivation to persist through the difficult times.

LESSON 5: TRADEOFFS BETWEEN TEACHING AND RESEARCH, OR NOT?

In the stories shared in this book, there was some tension between those that felt that they needed to decide between being a good teacher and a good researcher with others feeling that your teaching can inform your research and vice versa. Both Matthew and Charlie discuss being intentional about finding a faculty position at an institution that valued teaching. Inherent in these stories is the idea that there is a tradeoff between teaching and research, you are either a good teacher or a good researcher, but not likely both. Fernanda pushes on this idea by saying that teaching and research can be symbiotic. In other words, teaching can help give ideas for research and research can help give ideas for teaching.

What could faculty roles become if we started integrating our teaching and research efforts? How can we have teaching inform our research and our research inform our teaching? How can we begin to see these not as roles that are pitted against one another, but rather with each as integral parts of our roles as faculty members? If we are in administrative roles, how can we value both teaching and research in a way that encourages faculty to integrate these two aspects of their role.

LESSON 6: CONSIDER AN ASSET-BASED APPROACH TO YOUR TEACHING

An asset-based approach to teaching is one in which students are seen as having strengths and prior knowledge when they come into a classroom [Llopart and Esteban-Guitart, 2018]. They are seen as individuals with a myriad of experiences that will add to the class. In my experience, this approach is critical in improving our engineering education systems because many engineering faculty take a deficit-based approach, where they believe that students are not "cut-out" for engineering and lack the prerequisite knowledge (from K-12 or prior engineering classes) to do well in the class [Secules et al., 2018]. I have had many conversations in curriculum committees, faculty meetings, and hallways where faculty explain to me that our students are not smart enough to be engineering students, that our students are not adequately prepared to be engineering students, or that our students should be weeded out of their engineering programs. It was refreshing to read through stories of engineering faculty who do not take this deficit-based approach, but instead take an asset-based approach.

Fernanda discussed incorporating real-world projects into her courses. In one example, she describes asking questions in class and 1/3 of the student's hands go up immediately. The students have learned that she is interested in their particular experiences and know that the classroom is a space where they can share those experiences. She talks about how engaged they become, how well they can communicate when discussing something they know so well (their experiences), and how they seem to be more confident.

Also, Chris discussed his Site Remediation Techniques course that he transitioned to include real-world projects. The students became very engaged in the project as the projects involved real stakeholders in a real community. His students were distributed in this class with some having substantial practical experience and weaker technical backgrounds while others had strong technical backgrounds and limited practical experience. The students with practical experience were encouraged to bring their experience into the classroom to help teach the students with little practical experience about real-world engineering. This is different than many engineering classes, where technical knowledge is valued more than practical experience.

Donna took an asset-based approach in a deliberate way as she incorporated a liberative pedagogy in her classroom where she worked to change the power imbalance in the classroom. She adopted several strategies to give students agency and power in the class, primarily by having students critically reflect and critique the class.

How can we incorporate this lesson into our classrooms? For one, we can start asking explicitly about students' experiences as they relate to the course. In some courses it may be easy to value students' experiences, but in others it may appear more difficult at first glance. First, we can think about who has power in our classrooms and consider ways of giving the students agency within our classes. We could try having students read a handout by Foucalt or we can think about how to integrate real-world projects that build on experiences students bring into the classroom.

LESSON 7: EMPOWER ENGINEERING STUDENTS WHO HAVE OTHERWISE BEEN MARGINALIZED

Another transformative lesson emerged around the ability to empower engineering students who have traditionally been marginalized through our teaching approaches and Matthew was an exemplar of this. Early on in his own academic pursuits, Matthew befriended a guy who had ambitious goals with humble beginnings who would later become his best friend. His friend started as an auto mechanic attending community college and struggling academically. However, his desire was to land at NASA as a Rocket Scientist and with persistence and resilience, that is exactly where he ended up. This relationship was Matthew's first impactful exposure enabling him to recognize that not everyone had access to the same opportunities and access, and yet, where and/or how one started off on their journeys did not limit how far they could climb. Matthew realized that with the right support, encouragement and confidence, students could have a greater potential of achieving such laudable goals—inclusive of the "raw and scrappy" talent and not just those sought-after students that arrived prepared and ready to soar.

Matthew acknowledged that many of these students had not had much experience with college or positive experiences with the education system. He made a conscious decision that through his teaching, he would make a way for these students—the atypical recruit for engineering—to be able to see themselves as engineers. Through his openness and the camaraderie created through his teaching approaches, the students would be empowered "to ask questions maybe they had been afraid to ask before." The students were representations of his best friend and he chose to use his position to support them by meeting them where they were to help them develop and realize their fullest potential. It was important for Matthew that students learn "there's no stereotypical type of person that needs to be in this classroom."

The other experience that solidified the necessity of becoming an advocate for the marginalized engineering student occurred when Matthew changed his own last name. He took the last name of his wife, which happened to be a name of Hispanic origin. With this change, his identity of decades was suddenly challenged and he questioned, "I mean like now do I become a Hispanic engineer instead of an engineer?" Through the microaggressions that started to become commonplace for Matthew after changing his last name, he gained a different understanding of what "people deal with on a day-to-day basis" in terms of being underrepresented in engineering, and in society, in general. This shift in experience for a "stereotypical, white male in engineering" created a level of awareness that made Matthew want to create a space for dialogue, transparency and real conversations. Matthew reflected in his heightened awareness and position for advocacy, "But, I can't imagine going through the educational system and having to deal with that and learning new things all at the same time. I guess to bring it all back to the classroom environment, it's that kind of conversation that I want to have happen in engineering classrooms, because I think those are important conversations along with the technical."

How can we strive to have an awareness of the experiences of all students in our classrooms and empower those who have been marginalized in engineering? We learn through Matthew's story that there are several tangible actions we can take that can stand to have a significant impact on the students we encounter. The most critical of those being a self-reflection and acknowledgment of our own privilege and position (i.e., race, gender, education, socioeconomic status). We should challenge ourselves to use our position to push back against the systemic barriers facing students every day rather than being an added barrier to their load. We can also be open about our own identity with our students and unapologetic about who we are as Matthew exemplified, "I think they appreciate maybe that I'm just honest with who I am. At least that's my hope, that they recognize that you don't have to be afraid to be an individual, and you don't have to be afraid to learn." The last suggestion in fostering empowerment in the classroom is to encourage real conversations. Matthew urged that this didn't happen enough and people just "don't ask about how certain actions or things might make someone uncomfortable or something of the sort."

We can learn a lot from Matthew's example. What seems most encouraging is knowing that there is no perfect approach, it just takes a true desire and commitment to making a difference. Matthew was focused on helping students develop and what he envisioned as outcomes. At the end of his narrative, he describes a recent meeting with former students of his from his first transitional college. Most compelling is Matthew clearly remembers that perceiving these particular students through a deficit lens at a snapshot in time could have easily rendered, "Yeah, they're not going to be successful" for a lack of demonstrating traditional metrics of success. However, the students came to him—one finishing a Ph.D. in civil engineering and another a Master's in mechanical engineering—both of which most would regard as extremely successful. We cannot stop at how students show up in our classrooms, but as exemplar educators, must challenge ourselves to identify ways to empower them beyond the barriers to reach their fullest potential. It is our jobs as educators to help every student envision themselves as the engineer they wish to be. We are grateful for educators like Matthew and hope that others can be inspired and learn from his insight.

LESSON 8: CONNECTING THEORY TO THE REAL WORLD IN THE CLASSROOM

In some of the stories, there was a focus on providing opportunities for students to experience engineering practice. In many engineering classrooms, there tends to be a focus on technical solutions only, with no consideration of the complexity of solutions, especially when embedded in our social systems. Fernanda discusses a need to understand project complexity as we tend to over-simplify problems in engineering classes. She pushes students to think about "How do you connect the concepts that we covered in class to that real-world problem? How do you do something that people in the real world, e.g., a safety engineer, are actually doing?" To do this, she has strong connections with industry and includes real-world projects in her classes.

Chris also brings socio-technical problems into his classroom. For example, when he has students do watershed analysis he embeds that discussion with the impacts that watersheds can have on communities. For example, he discussed the Aberjona River watershed, as it is an example where there was contamination of a community's water which led to increased cases of leukemia. While this case is well known, it is also a case that is less than 20 miles from his university. Chris helps students make "connections between the abstract concept of watershed analysis and the concrete reality of understanding a watershed so that you can see its impacts on the community."

What are ways that we can make more explicit connections between the course curricula and our surrounding community? How can we begin to embed engineering problems into the real world? How can we begin to focus on the assumptions that we are making to teach engineering science courses? In research done by Erin Cech [2014], she found that students become more disengaged as they continue in an engineering program. Specifically, students' concerns about public welfare diminish as they continue in their studies. Through an attempt to make more connections between engineering and design and the communities that could be impacted by those designs, we could shift the paradigm to graduate engineers who are more engaged in public welfare.

LESSON 9: USING IDEAS FROM ENTREPRENEURSHIP IN ENGINEERING EDUCATION

Throughout many of the faculty stories, there was discussion of an entrepreneurial mindset helping teaching. Chris explains that to him, taking an entrepreneurial approach means thinking about the values of your students and developing ways of satisfying their values. Thais also explains that she had an "aha" moment when she realized that the students are her clients "and if they are not happy or if this is not useful for them, I have to do something." She then began attending Center for Teaching and Learning meetings and learning different teaching approaches so that she could begin to serve "more of the students and tailor my teaching to their needs."

Donna also encourages other engineering educators us to be entrepreneurial in her advice. She explains that you will always have constraints as you are trying to do things differently and push the boundaries, and advises you to "work creatively with, around, and through" those constraints. She explains that this creative "attitude" will help you continue to push boundaries and do things differently. Donna explains that "there are unchallenged assumptions everywhere" and that as innovative engineering educators, we have to begin pushing against those assumptions to truly be innovative. She also advises engineering educators to not be surprised when you encounter pushback from your peers, students, or administrators. This may actually be a sign that you are doing something right.

In Chris' story, he poses some questions for us to consider as educators,

> Why not be entrepreneurial in applying an educational concept? An educational innovation? When most of the education that we still receive today is the traditional lecture style, when people can deliver it in a different way, why can't that be an entrepreneurial effort?

Maybe considering ways of being more entrepreneurial or innovative in our approaches to teaching could help us become stronger teachers. When we are trying something new and it does not seem to be working well, when should we pivot to something new? When we are truly proposing something transformative in our teaching, should we expect to get some pushback from colleagues and students? These ideas of value propositions, customer segments, and pivots could be helpful as we begin pushing the boundaries of traditional engineering teaching and learning to do something truly innovative as engineering educators.

LESSON 10: COMFORT WITH AMBIGUITY AND RELINQUISHING CONTROL ARE REQUIRED

As faculty members, one natural tendency is to aim to cover all of the material as designated by the course syllabus. However, in the implementation of student-centered pedagogies, and particularly active learning, the proposed presentation of content does not always execute according to plan and this was one of the adjustments that the faculty in these stories had to get used to. These approaches necessitate a flexibility that challenges the certainty of knowing—whether it's the faculty's confidence in adequately covering the material or knowing exactly which topics the students will walk away from the class having mastered—there is a comfort with ambiguity and a relinquishing of control that essentially has to happen for faculty to shift their teaching.

This was demonstrated in Thais' story when she described the lack of a consistent expectation, "I never have the same exact lesson. Never, ever." For many faculty, the thought of such variation would be daunting. In fact, one of the reasons we invest such time and effort into developing a given lesson is the notion that the material will be repeatable and reusable. Brad mentioned that lecturing would simply be easier and that if he was solely focused on self, lecturing would be the rational choice. Thais acknowledged this tension when she admitted, "It's very hard to come to grips with the idea that you want to introduce these things and you want to give freedom for them to lead the class, and at the same time cover the course material."

How can we can as faculty learn to be open to the ambiguity and surrendering of control that is required when the outcomes of teaching approaches are less predictable? As we learn from Thais and Brad, having the confidence to try new things is imperative. Additionally, another way to navigate this struggle involves simply learning to accept that adopting active learning strategies may come with tradeoffs and/or require compromises. Students may not cover every single item in the course as they have historically, but the hope is that they will leave the class with a richer experience of engagement through thinking critically that fills those gaps all the same. In the words of Brad, "The workload is immensely higher than traditional teaching, but I think, just from my standpoint, I really see huge benefits to the students."

LESSON 11: LEARN SOMETHING NEW

A possible way to become better teachers is to become learners ourselves. In many of the stories, the faculty discussed becoming inspired when they were a student themselves. As a child, Sara knew she wanted to be a teacher. She had positive experiences as she pursued her undergraduate degree at Dartmouth, where she also served as a Teaching Assistant and experienced engineering faculty who cared about teaching. When she began to pursue her Ph.D. at a research-focused university, she was shocked by the poor quality of teaching she was experiencing. This, in part, motivated her even further to become a good teacher herself. Donna was also inspired, not when taking her engineering courses, but when taking humanities and social science courses as an undergraduate student. She discusses faculty from non-engineering departments that created an environment where she felt that she had something important to say, even though she had not taken some of the prerequisites for the course.

While many of these examples are of faculty becoming initially inspired to become better teachers through their experiences as students themselves, it makes me wonder if we can emulate those inspirations as the time between being students ourselves increases. Brad discusses that he enjoys learning and trying new things, and maybe we can take some inspiration from Brad. We can continue to learn new things and experience being a student throughout our lives. Recently, I began to learn to play the guitar. For me, this has been a huge inspiration as I get to experience first-hand being a complete novice and having so much to learn. In trying different ways of learning guitar, I have experienced different types of instructors. Some instructors have a fixed mindset and believe that either you can be a guitar player or you cannot, that somehow some people are born good guitar players. Another instructor who teaches online, goes to great lengths to explain that anyone can learn to play the guitar. He explains that babies are not born as guitar players and that everyone has to work at it. While it may come easier to some people, some of the greatest guitar players worked very hard to become the guitar players that they are today. Teaching using this growth mindset is an inspiration and helps you feel like you belong and that you, too, can become a guitar player. This has inspired me to change the way I talk about learning

Statics in one of my classes. Using this growth mindset explicitly in the class may help many of my students who were not as prepared academically as others in the class. Students need to be empowered and not held to the limitations of their preparation.

How can we continue to be inspired in our teaching? Maybe one way is to become a student ourselves. What is something that you've always wanted to learn, but never found the time? You could even extend the challenge to learn something completely different than your background—possibly learning to dance or paint. Maybe in learning something new you can become more inspired to become a stronger engineering educator. Maybe you can also begin to have more empathy for students who are struggling in your classes or those whose backgrounds have not equipped them with the tools of success.

CONCLUSION

Hopefully sharing the raw and real stories of engineering faculty in their transformations in teaching has inspired you in your own personal journey toward becoming an exemplary engineering educator. This book can serve as a catalyst for you to begin learning about others' teaching stories. Many of us have worked towards becoming better teachers, have encountered obstacles, while also experiencing some success. Continue these conversations by asking engineering educators about their stories of change and sharing your own journey.

REFERENCES

Cech, E. A. (2014). Culture of disengagement in engineering education? *Science, Technology, and Human Values*, 39(1), pp. 42–72. DOI: 10.1177/0162243913504305. 97

Llopart, M. and Esteban-Guitart, M. (2018). Funds of knowledge in 21st century societies: Inclusive educational practices for under-represented students. A literature review. *Journal of Curriculum Studies*. DOI: 10.1080/00220272.2016.1247913. 94

Secules, S., Gupta, A., Elby, A., and Turpen, C. (2018). Zooming out from the struggling individual student: An account of the cultural construction of engineering ability in an undergraduate programming class. *Journal of Engineering Education*. DOI: 10.1002/jee.20191. 94

Authors' Biographies (in order of appearance)

NADIA KELLAM

Nadia Kellam is an Associate Professor in the Polytechnic School of the Ira A. Fulton Schools of Engineering at Arizona State University. She is a qualitative researcher who primarily uses narrative research methods. In her research, Dr. Kellam is broadly interested in developing critical understandings of the culture of engineering education and, especially, the experiences of underrepresented undergraduate engineering students and engineering educators. In addition to teaching undergraduate engineering courses and a graduate course on entrepreneurship, she also enjoys teaching qualitative research methods in engineering education in the Engineering Education Systems and Design Ph.D. program at ASU. Nadia serves as Deputy Editor of the *Journal of Engineering Education*.

BROOKE COLEY

Brooke Coley is an Assistant Professor in Engineering at the Polytechnic School of the Ira A. Fulton Schools of Engineering at Arizona State University. Dr. Coley is Principal Investigator of the Shifting Perceptions, Attitudes and Cultures in Engineering (SPACE) Lab that aspires to elevate the experiences of marginalized populations, dismantle systematic injustices, and transform the way inclusion is cultivated in engineering through the implementation of novel technologies and methodologies in engineering education. Intrigued by the intersections of engineering education, mental health, and social justice, Dr. Coley's primary research interest focuses on virtual reality as a tool for developing empathetic and inclusive mindsets among engineering faculty. She is also interested in hidden populations in engineering education and innovation for more inclusive pedagogies.

AUDREY BOKLAGE

Audrey Boklage is a Research Assistant in the Center for Engineering Education of the Cockrell School of Engineering at The University of Texas at Austin. Prior to entering graduate school, she taught high school science in Texas for seven years. During this time, she redesigned curriculum and served as a mentor for new to profession educators. Upon receiving her doctorate degree in Curriculum and Instruction with a focus on STEM education, she became specifically interested in narrative research methods and faculty development within schools of engineering. Her current research interests include creating inclusive spaces within university engineering environments, specifically makerspaces and asset-based pedagogies.

DONA RILEY

Donna Riley is Kamyar Haghighi Head of the School of Engineering Education and Professor of Engineering Education at Purdue University. She is the author of two books, *Engineering and Social Justice* and *Engineering Thermodynamics and 21st Century Energy Problems,* both published by Morgan & Claypool. Riley earned a B.S.E. in Chemical Engineering from Princeton University and a Ph.D. from Carnegie Mellon University in Engineering and Public Policy. She is a fellow of the American Society for Engineering Education.

SARA ATWOOD

Sara Atwood is an Associate Professor and Chair of Engineering and Physics at Elizabethtown College. She received a B.A. and M.S. in Engineering Sciences from Dartmouth College and a Ph.D. in Mechanical Engineering from the University of California at Berkeley. She is passionate about engaging underrepresented students in engineering education, teaching engineers in a liberal arts setting, and encouraging students to use their engineering skills to be empowered citizens.

BRAD HYATT

Brad Hyatt is an Associate Professor and the Chair of the Department of Construction Management in the Lyles College of Engineering at California State University, Fresno (Fresno State). He has an M.S. in Engineering with a focus on Construction Engineering and Project Management from The University of Texas at Austin and a B.S. in Civil Engineering from the University of Kentucky. He teaches courses in construction estimating, scheduling, documents, and project controls. He actively conducts research on data and predictive analytics in construction, leadership in construction, lean construction practices, and integrating technology into construction pedagogy. Professor Hyatt continuously participates in leadership roles at Fresno State. He is a DISCOVERe Faculty Fellow and serves on the steering committee for the President's Leadership Academy. These transformational programs provide innovative solutions to mobile technology in the classroom and to the development of future leaders at Fresno State. Additionally, Professor Hyatt led a group of faculty to review learning management systems for the campus during the 2017/2018 academic year. Professor Hyatt is a Registered Professional Engineer in California and LEED Accredited Professional (Building, Design & Construction) with over 20 years of professional experience in program and project management of facilities, engineering, and construction projects. Professor Hyatt spent nearly ten years as a U.S. Navy Civil Engineer Corps Officer prior to his academic career. He also worked as a Construction Project Management consultant in between his military service and academic career. His broad industry expertise includes sustainable design and construction, facilities management, construction management, capital improvements planning, energy management, disaster response, and construction workforce shaping. He has managed a variety of projects from a large, complex replacement hospital to small fuel tank renovations.

CHRIS SWAN

Chris Swan is Dean of Undergraduate Education in the School of Engineering at Tufts University and an Associate Professor in its Civil and Environmental Engineering Department. He is also a senior fellow in Tisch College of Civic Life. Previously, he has served as CEE department chair. He received a ScD degree in Civil and Environmental Engineering from MIT in 1994 and both B.S. and M.S. degrees in Civil Engineering from the University of Texas at Austin in 1984 and 1986, respectively. An initiator of explicitly incorporating components of service-learning into engineering curriculum at Tufts, he continues to champion the development and implementation of civic engagement in engineering education. For example, he currently serves as an advisor to Tufts student chapter of Engineers Without Borders. Current engineering education research efforts focus on evaluating the impact of service-based learning in engineering education, as well as applying entrepreneurial principles in examining sustainable and scalable pathways for innovations in engineering education. He was also an inaugural Faculty Fellow of Tisch College and of the Center for the Enhancement of Learning and Teaching (CELT). In addition, Chris researches the development of reuse strategies for waste materials. Most notably, his research efforts have focused on the reuse of fly ash from coal burning facilities with waste plastics. This has led to the development of synthetic lightweight aggregates (SLA), an innovative construction material that can be used in place of traditional sand and gravel.

THAIS ALVES

Thais Alves specializes in construction management and project-based production systems. Her areas of interest include the application of Lean production/construction concepts, principles, and tools to improve the performance of production systems and products in different stages of their life-cycle and supply chains. Additionally, she is interested in how contracts and delivery methods support collaboration across supply chains in the Owner-Architecture-Engineering-Construction industry. For more than 15 years, Thais has been teaching, advising students, researching, and collaborating with construction companies toward the dissemination and implementation of Lean, especially in the field of production planning and control at construction sites. She is currently the AGC—Paul S. Roel Chair in Construction Engineering and Management at the J.R. Filanc Construction Engineering and Management Program at San Diego State University.

FERNANDA LEITE

Fernanda Leite is an Associate Professor in Construction Engineering and Project Management, in the Civil, Architectural and Environmental Engineering (CAEE) Department at the University of Texas at Austin. She has a Ph.D. in Civil and Environmental Engineering from Carnegie Mellon University. Prior to her graduate education, she worked as a Project Manager in her home country of Brazil, in multiple government infrastructure and commercial building construction projects. Her technical interests include information technology for project management, building information modeling, collaboration and coordination technologies, and information technology-supported construction safety management. She has taught four unique courses at UT and has integrated project-based and experiential learning to all of her courses, through class projects, industry mentorships, and interactive exercises. She serves as Graduate Program Coordinator for CAEE's Sustainable Systems graduate program and on the Executive Committee for the University-wide Grand Challenges effort called Planet Texas 2050. She currently supervises 8 Ph.D. and 5 M.S. students. She has graduated 7 Ph.D. and 36 M.S. students.

CHARLES E. PIERCE

Charles E. Pierce is an Associate Professor in the Department of Civil and Environmental Engineering at the University of South Carolina (USC), where he has been teaching since 1998. He has an M.S. and Ph.D. in Civil Engineering from Northwestern University and a B.S. degree in Civil Engineering from the University of New Hampshire. He is the current Director for Diversity and Inclusion in his department and a USC Connect Faculty Fellow for Integrative Learning. He was awarded the Michael J. Mungo Undergraduate Teaching Award for USC in 2006, and he is also the recipient of the Samuel P. Litman Award and Bell South Teaching Fellowship in recognition of his contributions to engineering education. Dr. Pierce is an active member of ASEE and serves as the campus representative for USC. He is committed to improving engineering education across the K-20 spectrum. His contributions include leading professional development activities on engineering for middle and high school math and science teachers and creating programs for graduate students in engineering to integrate research and teaching. His undergraduate educational interests include the facilitation and assessment of critical thinking through problem-based learning using the Environments for Fostering Effective Critical Thinking (EFFECTs) framework developed with his colleagues at USC.

MATTHEW FUENTES

Matthew Fuentes is currently a member of the engineering faculty at Everett Community College. He has been teaching at community colleges for 10 years. He earned a B.S. and M.S. in Aerospace Engineering from the University of Tennessee.

Printed in the United States
by Baker & Taylor Publisher Services